Lecture Notes in Computer Science 12969

More information about this subseries at http://www.springer.com/series/7412

Cristina Oyarzun Laura · M. Jorge Cardoso ·
Michal Rosen-Zvi · Georgios Kaissis et al. (Eds.)

Clinical Image-Based Procedures, Distributed and Collaborative Learning, Artificial Intelligence for Combating COVID-19 and Secure and Privacy-Preserving Machine Learning

10th Workshop, CLIP 2021
Second Workshop, DCL 2021, First Workshop, LL-COVID19 2021
and First Workshop and Tutorial, PPML 2021
Held in Conjunction with MICCAI 2021
Strasbourg, France, September 27 and October 1, 2021
Proceedings

Editors
Cristina Oyarzun Laura
Fraunhofer IGD
Darmstadt, Germany

M. Jorge Cardoso
King's College London
London, UK

Michal Rosen-Zvi
IBM Research-Haifa and Haifa University
Haifa, Israel

Georgios Kaissis
Technical University of Munich
Munich, Germany

Additional Editors *see next page*

ISSN 0302-9743 ISSN 1611-3349 (electronic)
Lecture Notes in Computer Science
ISBN 978-3-030-90873-7 ISBN 978-3-030-90874-4 (eBook)
https://doi.org/10.1007/978-3-030-90874-4

LNCS Sublibrary: SL6 – Image Processing, Computer Vision, Pattern Recognition, and Graphics

This Springer imprint is published by the registered company Springer Nature Switzerland AG
The registered company address is: Gewerbestrasse 11, 6330 Cham, Switzerland

Additional Editors

Marius George Linguraru
Children's National Health System
Washington, D.C., DC, USA

Raj Shekhar
Children's National Health System
Washington, DC, USA

Stefan Wesarg
Fraunhofer IGD
Darmstadt, Germany

Marius Erdt
Fraunhofer Singapore
Singapore, Singapore

Klaus Drechsler
Aachen University of Applied Sciences
Jülich, Germany

Yufei Chen
Tongji University
Shanghai, China

Shadi Albarqouni
Helmholtz AI
Neuherberg, Germany

Spyridon Bakas
University of Pennsylvania
Philadelphia, PA, USA

Bennett Landman
Vanderbilt University
Nashville, TN, USA

Nicola Rieke
NVIDIA GmbH
Munich, Germany

Holger Roth
NVIDIA Corporation
Bethesda, MD, USA

Xiaoxiao Li
University of British Columbia
Vancouver, Canada

Daguang Xu
NVIDIA Corporation
Santa Clara, CA, USA

Maria Gabrani
IBM Research Europe
Rueschlikon, Switzerland

Ender Konukoglu
Computer Vision Laboratory
Zürich, Switzerland

Michal Guindy
Assuta Medical Centers Radiology
Aviv-Yafo, Israel

Daniel Rueckert
Technical University of Munich
Munich, Germany

Alexander Ziller
Technical University of Munich
Munich, Germany

Dmitrii Usynin
Technical University of Munich
Munich, Germany

Jonathan Passerat-Palmbach
Imperial College London
London, UK

CLIP Preface

The 10th International Workshop on Clinical Image-Based Procedures: Towards Holistic Patient Models for Personalised Healthcare (CLIP) was held on September 27, 2021, in conjunction with the 24th International Conference on Medical Image Computing and Computer Assisted Intervention (MICCAI 2021). As with the previous edition, CLIP 2021 was held virtually to keep all participants safe from the current COVID-19 pandemic while facilitating researchers to participate in spite of the strict traveling rules.

Following the long tradition of CLIP on translational research, the goal of the works presented in this workshop is to bring basic research methods closer to the clinical practice. One of the key aspects that is gaining relevance regarding the applicability of basic research methods in clinical practice is the creation of Holistic Patient Models as an important step towards personalized healthcare. As a matter of fact, the clinical picture of a patient does not uniquely consist of medical images, instead a combination of medical image data of multiple modalities with other patient data, e.g. omics, demographics, or electronic health records, is desirable. Since 2019 CLIP has put a special emphasis on this area of research.

As in the previous CLIP workshops that have taken place every year since 2012, all submitted papers were peer reviewed by at least two experts. CLIP 2021 received 13 submissions and nine of them were accepted for publication. All accepted papers were presented by their authors during the workshop and the attendees chose with their votes the recipient of the Best Paper Award of CLIP 2021. In addition to the oral presentations provided by the authors of the accepted papers, all attendees of CLIP 2021 had the opportunity to enjoy high-quality keynotes followed by avid discussions in which all attendees were involved. We would like to thank our invited speakers for their interesting talks and discussions. Furthermore, we would like to take this opportunity to thank our Program Committee members, authors, and attendees who helped CLIP 2021 to be a great success.

September 2021

Cristina Oyarzun Laura
Marius George Linguraru
Raj Shekhar
Stefan Wesarg
Marius Erdt
Klaus Drechsler
Yufei Chen

CLIP Preface

[illegible]

[illegible]

[illegible]

[illegible]

[illegible]

CLIP Organization

Program Chairs

Yufei Chen	Tongji University, China
Klaus Drechsler	Aachen University of Applied Sciences, Germany
Marius Erdt	Fraunhofer Singapore, Singapore
Marius George Linguraru	Children's National Health System, USA
Cristina Oyarzun Laura	Fraunhofer IGD, Germany
Raj Shekhar	Children's National Health System, USA
Stefan Wesarg	Fraunhofer IGD, Germany

Program Committee

Xin Kang	Sonavex, USA
Jan Egger	TU Graz, Austria
Roman Martel	Fraunhofer Singapore, Singapore
Juan Cerrolaza	QUANT AI Lab, Spain
Yogesh Karpate	Childrens National Medical Center, USA
Katarzyna Heryan	AGH UST, Poland
Chaoqun Dong	Fraunhofer Singapore, Singapore
Stephan Zidowitz	Fraunhofer MEVIS, Germany
Weimin Huang	Institute for Infocomm Research, Singapore
Xingzi Zhang	NTU, Singapore

DCL Preface

Machine learning approaches have demonstrated the capability of revolutionizing almost every application and every industry through the use of large amounts of data to capture and recognize patterns. A central topic in recent scientific debates has been how data is obtained and how it can be used without compromising user privacy. Industrial exploitation of machine learning and deep learning (DL) approaches has, on the one hand, highlighted the need to capture user data from the field of application in order to yield a continuous improvement of the model, and on the other hand it has exposed a few shortcomings of current methods when it comes to privacy.

Innovation in the way data is captured, used, and managed, as well as how privacy and security of this data can be ensured, is a priority for the whole research community. Most current methods rely on centralized data stores, which contain sensitive information and are often out of the direct control of users. In sensitive contexts, such as healthcare, where privacy takes priority over functionality, approaches that require centralized data lakes containing user data are far from ideal, and may result in severe limitations in what kinds of models can be developed and what applications can be served.

Other issues that result in privacy concerns are more intimately connected with the mathematical framework of machine learning approaches and, in particular, DL methods. It has been shown that DL models tend to memorize parts of the training data and, potentially, sensitive information within their parameters. Recent research is actively seeking ways to reduce issues caused by this phenomenon. Even though these topics extend beyond distributed and collaborative learning methods, they are still intimately connected to them.

The second MICCAI workshop on Distributed and Collaborative Learning (DCL 2021) aimed at creating a scientific discussion focusing on the comparison, evaluation, and discussion of methodological advancement and practical ideas about machine learning applied to problems where data cannot be stored in centralized databases; where information privacy is a priority; where it is necessary to deliver strong guarantees on the amount and nature of private information that may be revealed by the model as a result of training; and where it's necessary to orchestrate, manage, and direct clusters of nodes participating in the spotential conflicts of interest and recent ame learning task.

During the second edition of DCL, eight papers were submitted for consideration and, after peer review, four full papers were accepted for presentation. Each paper was rigorously reviewed by at least three reviewers in a double-blind review process. The papers were assigned to reviewers taking into account (and avoiding) potential conflicts of interest and recent work collaborations between peers. Reviewers were selected from among the most prominent experts in the field from all over the world.

Once the reviews were obtained, the area chairs formulated final decisions over acceptance, conditional acceptance, or rejection of each manuscript. These decisions were always taken according to the reviews and could not be appealed. In the case of conditional acceptance, authors had to make substantial changes and improvements to their paper according to reviewer feedback. The nature of these changes aimed to increase the scientific validity as well as the clarity of the manuscripts.

Additionally, the workshop organizing committee granted the Best Paper Award to the best submission presented at DCL 2021. The Best Paper Award was assigned as a result of a secret voting procedure where each member of the committee indicated two papers worthy of consideration for the award. The paper collecting the majority of votes was then chosen by the committee.

The double-blind review process with three independent reviewers selected for each paper, united with the mechanism of conditional acceptance, as well as the selection and decision process through meta-reviewers, ensured the scientific validity and the high quality of the works presented at the second edition of DCL, making our contribution very valuable to the MICCAI community, and in particular to researchers working on distributed and collaborative learning.

We would therefore like to thank the authors for their contributions, and the reviewers for their dedication and fairness when judging the works of their peers.

August 2021

Shadi Albarqouni
Spyridon Bakas
M. Jorge Cardoso
Bennett Landman
Xiaoxiao Li
Nicola Rieke
Holger Roth
Daguang Xu

DCL Organization

Organizing Committee

Shadi Albarqouni	Helmholtz AI and Technical University of Munich, Germany
M. Jorge Cardoso	King's College London, UK
Nicola Rieke	NVIDIA, Germany
Daguang Xu	NVIDIA, USA
Spyridon Bakas	University of Pennsylvania, USA
Bennett Landman	Vanderbilt University, USA
Xiaoxiao Li	University of British Columbia, Canada
Holger Roth	NVIDIA, USA

Area Chairs

Shadi Albarqouni	Helmholtz AI and Technical University of Munich, Germany
M. Jorge Cardoso	King's College London, UK
Spyridon Bakas	University of Pennsylvania, USA
Holger Roth	NVIDIA, USA

Program Committee

Aaron Carass	Johns Hopkins University, USA
Amir Alansary	Johns Hopkins University, USA
Andriy Myronenko	NVIDIA, USA
Benjamin A. Murray	King's College London, UK
Christian Wachinger	LMU Munich, Germany
Daniel Rubin	Stanford University, USA
Dong Yang	NVIDIA, USA
Ipek Oguz	Vanderbilt University, USA
G. Anthony Reina	Intel Corporation, USA
Jayashree Kalpathy-Cramer	Massachusetts General Hospital, USA
Jonas Scherer	DKFZ, Germany
Jonny Hancox	NVIDIA, UK
Kate Saenko	Boston University, USA
Ken Chang	Massachusetts General Hospital, USA
Khaled Younis	GE Healthcare, USA
Klaus Kades	DKFZ, Germany
Ling Shao	Inception Institute of Artificial Intelligence, Abu Dhabi, UAE
Marco Nolden	DKFZ, Germany

Maximilian Zenk	DKFZ, Germany
Meirui Jiang	Chinese University of Hong Kong, Hong Kong
Micah J. Sheller	Intel Corporation, USA
Nir Neumark	Massachusetts General Hospital, USA
Qiang Yang	HKUST, Hong Kong
Quande Liu	Chinese University of Hong Kong, Hong Kong
Quanzheng Li	Massachusetts General Hospital, USA
Ralf Floca	DKFZ, Germany
Sarthak Pati	University of Pennsylvania, UK
Shunxing Bao	Vanderbilt University, USA
Walter Hugo Lopez Pinaya	King's College London, UK
Wojciech Samek	Fraunhofer HHI, Germany
Xingchao Peng	Boston University, UK
Yang Liu	WeBank, China
Yuankai Huo	Vanderbilt University, USA
Zach Eaton-Rosen	King's College London, UK
Zijun Huang	Columbia University, USA
Ziyue Xu	NVIDIA, USA

LL-COVID-19 Preface

During the global COVID-19 pandemic we observed a global pressure on the healthcare systems that drove increased leverage of lung medical imaging for diagnosis, prognosis, and treatment selection. This resulted in a surge of publications exploring, from one side, clinical use of medical imaging by the COVID-19 patients' carers, and AI models analyzing lung medical images developed by the AI community, on the other side. Unfortunately, the publications demonstrating the use of AI models in clinical practice were rather limited. We performed a thorough review of all relevant publications in 2020 [1] and identified numerous trends and insights that may help in accelerating the translation of AI technology in clinical practice in pandemic times. Aiming to continue and expand the discussion, the LL-COVID-19 MICCAI 2021 workshop was devoted to the lessons learned from this accelerated process and in paving the way for further AI adoption, in particular focusing on three main areas, namely (1) data definition, (2) data availability, and (3) research translation, as we reason and describe in detail in the first paper of the volume.

The program of the LL-COVID-19 MICCAI workshop was designed to facilitate discussion between the AI community and medical experts. To this end, in each of our three sessions we included (1) presentations of peer-reviewed papers, (2) presentations from invited speakers to cover domains and expertise not covered by the accepted papers, and (3) panel discussions with all presenters and additional invited panelists to expand the expertise, demographics, and modalities.

To select the peer-reviewed papers we leveraged the CMT tool. We applied a double-blind review process and had each submitted paper reviewed by three independent reviewers. The invited speakers were selected based on their contributions to the COVID-19 publications, as identified in the publications review process we performed. The panelists were selected mostly based on geographic and modalities coverage. The invited speakers and additional panelists are listed on the organization page.

September 2021

Michal Rosen-Zvi
Maria Gabrani
Ender Konukoglu
David Beymer
Gustavo Carneiro
Michal Guindy

LL-COVID-19 Organization

General Chair

Michal Rosen-Zvi | IBM Research, Israel

Program Committee Chairs

Maria Gabrani	IBM Research - Zurich, Switzerland
Ender Konukoglu	ETH Zurich, Switzerland
David Beymer	IBM Research - Almaden, USA
Gustavo Carneiro	University of Adelaide, Australia
Michal Guindy	Assuta Medical Center, Israel

Additional Reviewers

Jannis Born
Nathaniel Braman
Colin Campas
Ehsan Dehghan
Davide Fontanarosa
Fengbei Liu
Henning Mueller
Anirban Mukhopadhyay
Jacinto Nascimento
Deepta Rajan

Invited Speakers

Janis Born	IBM Research - Zurich, Switzerland
Anirban Mukhopadhyay	Technical University of Darmstadt, Germany
Ruud van Sloun	Technical University Eindhoven, The Netherlands
Colin Campas	NVIDIA, USA
Holger Roth	NVIDIA, USA
Dorit Shaham	Hadassah Medical Center, Israel
Bram van Ginneken	Radboud University Medical Center, The Netherlands

Invited Panelists

Itamar Ofer	World Hospital at Home Community and Congress, Israel
Bishesh Khanal	NAAMII, Nepal
Orest Boyko	Indiana University Health, USA

Reference

1. Born, J., et al.: On the role of artificial intelligence in medical Imaging of COVID-19. Patterns, **2**(6), 1–18 (2021)

PPML Preface

Any handling of medical data, be it in the setting of research or for clinical patient care, by definition involves an interaction with privacy-sensitive attributes inherent to this data. Both the development of novel AI techniques and their validation and future application in healthcare will rely on training and testing of algorithms on large, multi-institutional data sets. This is crucial both for reasons of data representativeness and for fairness and unbiasedness. It becomes apparent that the aforementioned privacy constraints directly contradict the common paradigm of centralized data pooling enabled through data sharing agreements and *carte blanche* institutional review board approval. They infringe on data ownership, preclude the enforcement of granular data governance by creating identical copies of datasets and leave data vulnerable to theft or inappropriate handling. Data trusts provide only limited relief, as they still rely on data centralization. Instead, future-proof solutions based around technical means of privacy enforcement on the patient level and decentralized data utilization without direct data access are required. Privacy-preserving machine learning techniques provide concrete solutions: they enable the security, privacy, and verification of algorithms and data and allow the enforcement of data flow governance, i.e. the compliance of information flow to, for example, contractually agreed norms.

Concrete technical implementations of such systems rely on a novel and expanding set of technical implementations. The remote execution of data processing workflows is typically realized through techniques such as federated learning, in which data processing algorithms, such as deep neural networks, are dispatched to the site where the data resides and are returned to a central server after the data has been processed. Notably this step does not entail transmitting any of the actual patient data. Even in the remote execution setting, however, algorithms trained on sensitive datasets are potentially themselves a source of sensitive information. Reverse engineering of such models can relieve such information to malicious third parties. Objective privacy guarantees are provided by the conceptual and technical framework of differential privacy, which allows the quantification of individual privacy and the fine-grained regulation of the interaction between the dataset and the algorithm or the dataset and the data scientist through privacy budgets.

Our aim with the MICCAI 2021 PPML workshop was to empower the medical imaging community to explore these and other techniques (such as cryptography and verification). It is our conviction that research progress in our field will benefit from their application and their development in concrete research settings, as well as their eventual translation to clinical patient care.

September 2021

Georgios Kaissis
Alexander Ziller
Daniel Rueckert
Dmitrii Usynin
Jonathan Passerat-Palmbach

PPML Organization

General Chair

Georgios Kaissis	Technical University of Munich, Germany

Program Committee Chairs

Alexander Ziller	Technical University of Munich, Germany
Daniel Rueckert	Technical University of Munich, Germany
Dmitrii Usynin	Technical University of Munich, Germany
Jonathan Passerat-Palmbach	Imperial College London, UK

Program Committee

Shaistha Fathima	OpenMined, India
Abinav Ravi Venkatakrishnan	deepc, Germany
Gharib Gharibi	TripleBlind AI, USA
Moritz Knolle	Technical University of Munich, Germany
Fatemeh Mireshghallah	University of California, San Diego, USA
Ajinkya Mulay	Purdue University, USA
Reza Nasirigerdeh	Technical University of Munich, Germany
Tushar Semwal	University of Edinburgh, UK
Madeleine Shang	OpenMined, Canada
Reihaneh Torkazdehmahani	Technical University of Munich, Germany

Contents

CLIP

DCL

LL-COVID19

PPML

CLIP

Intestine Segmentation with Small Computational Cost for Diagnosis Assistance of Ileus and Intestinal Obstruction

Hirohisa Oda[1(✉)], Yuichiro Hayashi[2], Takayuki Kitasaka[3], Aitaro Takimoto[1], Akinari Hinoki[1], Hiroo Uchida[1], Kojiro Suzuki[4], Masahiro Oda[1,5], and Kensaku Mori[1,6,7]

[1] Nagoya University Graduate School of Medicine, Nagoya, Japan
hoda@mori.m.is.nagoya-u.ac.jp
[2] Graduate School of Informatics, Nagoya University, Nagoya, Japan
[3] School of Information Science, Aichi Institute of Technology, Toyota, Japan
[4] Department of Radiology, Aichi Medical University, Nagakute, Japan
[5] Strategy Office, Information and Communications, Nagoya University, Nagoya, Japan
[6] Information Technology Center, Nagoya University, Nagoya, Japan
[7] Research Center for Medical Bigdata, National Institute of Informatics, Chiyoda City, Japan

Abstract. This paper proposes an intestine segmentation method from CT volumes for the intestinal obstruction and the ileus diagnosis assistance. The previous method was built based on the 3D U-Net, whose computational cost was high. Nevertheless, there was no confirmation that the 3-dimensional network contributed to the segmentation performance. In this paper, we propose a method utilizing the 2D U-Net on behalf of the previous methods' 3D U-Net. Experimental results using 110 CT volumes showed that both the proposed (2D U-Net) and previous (3D U-Net) methods achieved similar scores for the segmentation accuracy and system usefulness. In addition, the proposed method's inference was 0.5 min on average, around 8-times faster than the 3D U-Net's. Although subsequent processes still require more than 20 min for each case, utilizing lightweight networks is essential for practical use in the future, especially for emergency diagnosis.

Keywords: Sparse annotation · FCN · Small bowel segmentation

1 Introduction

Intestinal obstruction and the ileus are diseases that prevent normal movement of the intestines' contents. Those diseases require image diagnosis using abdominal CT volumes. Clinicians observe abdominal CT volumes to find the diseased points that are obstructing or paralyzing. However, those diseased points often

C. Oyarzun Laura et al. (Eds.): CLIP/DCL/LL-COVID/PPML 2021, LNCS 12969, pp. 3–12, 2021.
https://doi.org/10.1007/978-3-030-90874-4_1

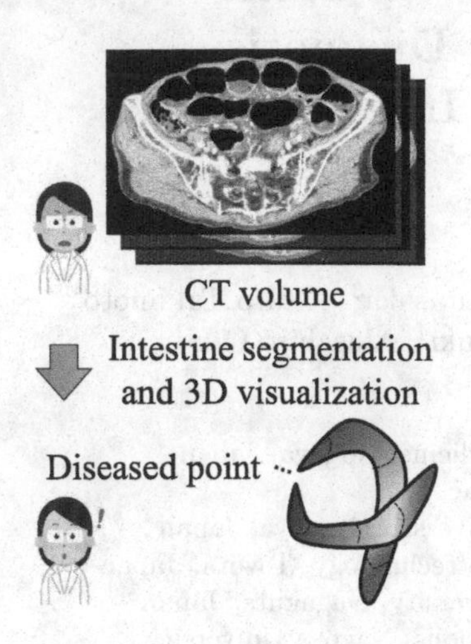

Fig. 1. Computer-aided detection (CADe) system design. Intestine segmentation allows us to visualize intestines intuitively. Endpoints of segmentation results might be diseased points that intestines' contents cannot move forward.

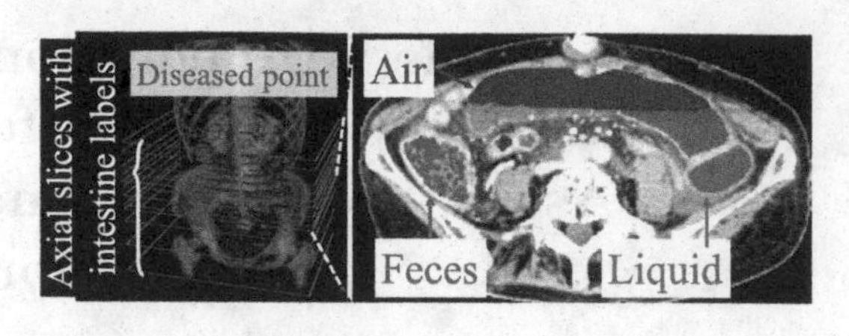

Fig. 2. Intestine labels annotated on several axial slices. Although the labels are divided into classes of intestine contents (air, liquid, and feces), these are all handled as binary labels by the methods. Diseased points are also annotated.

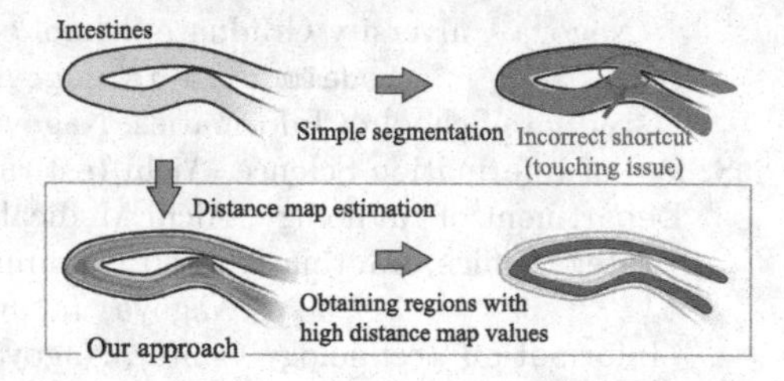

Fig. 3. Motivation of distance map estimation. Intestines are long and winding. Some parts are touching each other. Simple segmentation often generates incorrect shortcuts. Distance maps are firstly estimated, and then regions with high distance map values are obtained.

exist in the small intestines. Contrast agents are utilized for blood vessels, and fecal tagging is not available. Therefore, it is difficult for non-expert clinicians to imagine intestine structures and find the diseased parts because the small intestines are complicatedly winding [1]. Furthermore, this diagnosis is often performed on the emergency diagnosis, which requires diagnosing accurately and quickly.

For assisting accurate and quick diagnosis, computer-aided detection (CADe) systems are desired [2]. Figure 1 illustrates our system design, and such system's core is intestine segmentation. Unfortunately, there are very few works [3,4] for small intestines from CT volumes due to the complicated structures. A paper [5] proposed an intestine segmentation method. Their 3D visualization of segmentation results is intuitive for understanding the intestines' structures. Furthermore, since diseased points are usually thin, endpoints of segmentation results are often located near the diseased points. However, the previous method's significant problem is its computational cost. It takes around 30 min for intestine segmentation on each patient, even if a GPU is utilized. A computationally heavy part is the 3D U-Net. Especially for practical use in the emergency diagnosis, discovering more lightweight methods is essential.

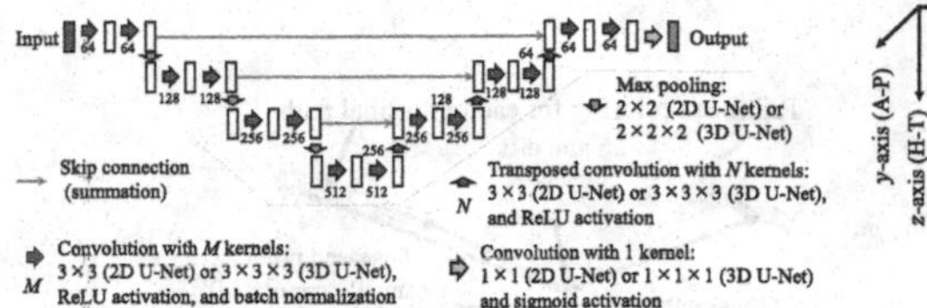

Fig. 4. Network structures of 2D and 3D U-Nets we implemented. Structures are almost common between two networks, except for dimensions of input, output, and layers.

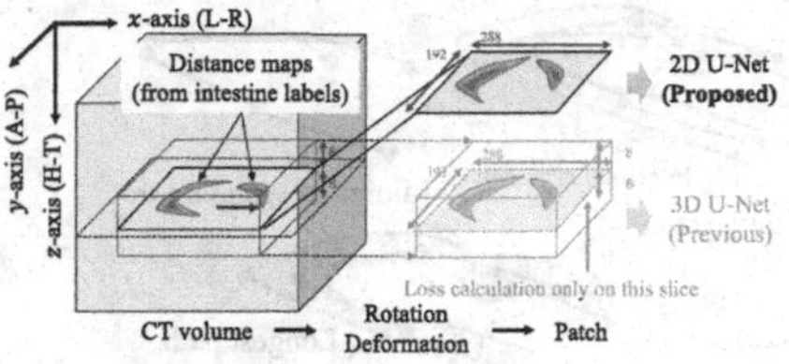

Fig. 5. Patch generation for training 2D or 3D U-Net. 2D or 3D patches around axial slices having intestine labels are cropped, for 2D or 3D U-Net, respectively. Rotation and deformation are performed for data augmentation.

On Oda, et al.'s scheme [5], only several axial slices contain manually-traced intestine labels because generating labels on all slices is tough for human laborers. The 3D U-Net [6] was utilized for considering several axial slices around the annotated slice during training. However, the 3D U-Net's efficacies for segmentation performance or system usefulness were not validated compared to the 2D U-Net [7]. The 3D U-Net's computational cost is much higher than those of the 2D U-Net. There is no apparent motivation to use the 3D U-Net.

In this paper, we propose an intestine segmentation method that simply utilizes the 2D U-Net. The 2D U-Net's training and inference are much faster than the 3D U-Net's. We compare the proposed method's segmentation accuracy and computational time to those of the 3D U-Net.

2 Methods

2.1 Overview

We propose an intestine segmentation method utilizing the 2D U-Net [7]. The previous method [5] used the 3D U-Net [6] for segmentation. Those networks' input and output are a CT volume and its intestine segmentation result, respectively. Training of the network is required before segmentation with the training dataset. Segmentation results are visualized so that suitable for diagnosis assistance of the intestinal obstruction and the ileus.

2.2 Distance Map Estimation for Preventing Incorrect Shortcuts

Motivation. The intestines, including the small intestines, are long and complicated winding in the abdomen. Therefore, different parts of the intestines are often touching each other (called "touching issue" in Shin, et al. [4]). In such a case, segmentation results should not have shortcuts between the touching parts.

Oda, et al. [5] introduced a distance map estimation for solving the touching issue, as illustrated in Fig. 3. Instead of training with the images and labels, the

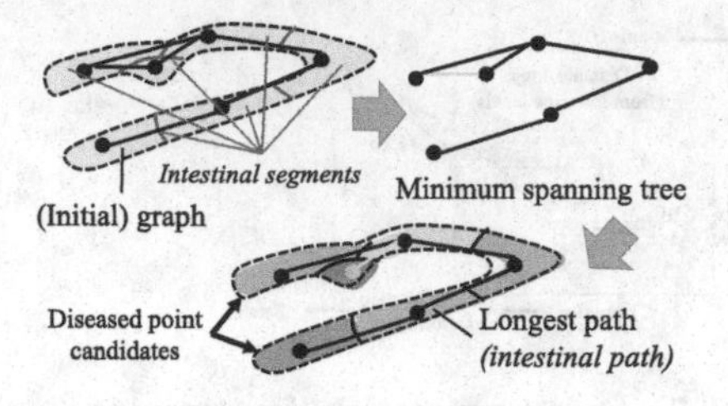

Fig. 6. Graphical process (Sect. 2.3). Graphs are initially generated for representing connections of intestinal segments. After converting to minimum spanning tree, longest path is obtained as intestinal path. Coloring is performed along intestinal path.

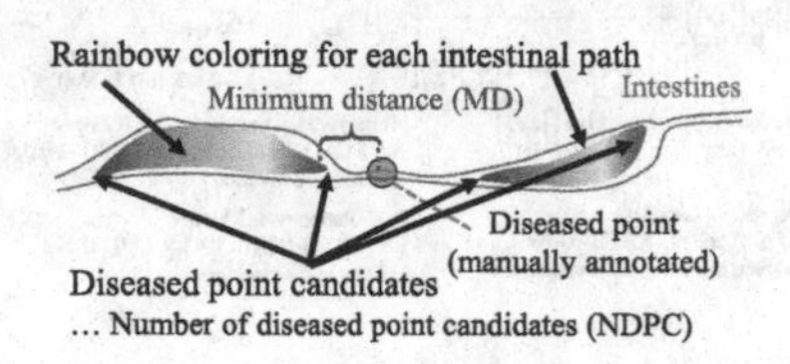

Fig. 7. Coloring of intestinal paths (Sect. 2.3) and metrics for system usefulness (Sect. 3.2).

labels are converted to distance maps and use them for training. The distance maps have the highest values (near 1.0) on the intestines' centerlines and lower in the peripheral or outside the intestines. The trained network can estimate distance maps on CT volumes for testing. Obtaining regions with high distance maps cover regions around intestines' centerlines, and incorrect shortcuts are now prevented.

Generating Distance Maps for Training Dataset. Each CT volume in the training dataset contains manually-traced labels of intestines (1 inside the intestines and 0 outside them), as shown in Fig. 2. Those labels exist only on several axial slices.

Although it is possible to train the FCNs using the labels binarized to 0 or 1, we convert the labels to the distance maps normalized into the range [0, 1] by the 2D distance transformation for each axial slice containing labels. The network is trained with pairs of CT volumes and distance maps. The networks and their usage will be described in Sect. 2.2, respectively.

Training and Inference. The networks are defined below. As illustrated in Fig. 4, we utilized those networks with similar settings, e.g., the number of layers or filters. The network structures are illustrated in Fig. 4.

- **2D U-Net (used in the proposed method)**
  The 2D U-Net [7] is a 2D FCN whose input and output are 2D patches. For intestine segmentation in this paper, we train it using pairs of axial slices (288 $\times$ 192 pixels) and their intestine labels. This is simpler than extracting 3D patches as well as the 3D U-Net described below. The number of trainable parameters is 7,530,113.
- **3D U-Net (used in** [5])
  The 3D U-Net [6] is a 3D FCN whose input and output are 3D patches. The network described in [5] has differences from the original paper [6] for training

with weakly-annotated intestine labels. Input and output patches consist of $288 \times 192 \times 16$ voxels. The 9th X-Y plane from the top contains the distance maps. The number of trainable parameters is 22,571,777.

The network is trained using patches generated from CT volumes and their intestine label on several axial slices, as illustrated in Fig. 5. A patch is generated from a randomly selected axial slice containing the intestine labels converted to the distance maps. Random rotation and non-rigid deformation are applied as data augmentation.

The inference is performed for all axial slices in CT volumes in the testing dataset, which estimates the distance maps. Regions with high distance map values are obtained as intestine regions.

2.3 Graph-Based Segmentation and Visualization

Graphical Representation with Watershed Algorithm. Thresholding against network output can directly produce intestine regions. However, it is important to visualize how the intestines are running inside the abdomen. Therefore, on behalf of thresholding, [5] introduced graphical representation. Although this part is not the core of this paper, the process of [5] is summarized below.

On the estimated distance maps, regions are obtained by the Watershed algorithms from local maxima (each result of the Watershed algorithm is called "intestinal segments") with the threshold t. The threshold t for the Watershed algorithm should be high to prevent incorrect shortcuts between different sections adjacent to each other. Each connected component between those intestinal segments is regarded as a graph (Fig. 6). Each intestinal segment is considered as a node, and contacts between intestinal segments are regarded as edges. The distance between two nodes (intestinal segments) are defined by the Euclid distance between their centers. Those graphs are converted to the minimum spanning trees. For each connected component, the longest path between leaf nodes is considered an "intestinal path" running through the intestines.

Visualization. For obtaining the intestine regions sufficiently until the peripheral parts, another threshold $s (0 < s < t < 1)$ is utilized. Also, suppose endpoints of two intestinal paths contact with the threshold s. In that case, those intestinal paths are merged to correct false divisions around the curves.

Coloring is performed along the intestinal path, as shown in Fig. 7. A rainbow-color scheme colors the intestinal segments which are corresponding to the nodes on the intestinal path. Also, endpoints of intestinal paths that might be diseased points are colored in red or blue. This visualization scheme allows clinicians to track intestines intuitively.

3 Experimental Results

3.1 Experimental Setup

We perform intestine segmentation on 110 CT volumes of the intestinal obstruction or the ileus patients. Those CT volumes are divided into four groups at the patient level for cross-validation. On the CT volumes, a trained medical student manually traced labels, as shown in Fig. 2. For each CT volume, the labels were traced on randomly selected 7–10 axial slices. Then, a surgeon checked those labels' correctness. Although the labels have information on intestine content types (air, liquid, or feces), this information is not utilized for training or evaluation. In addition, the surgeon annotated diseased points (only one for most cases) for CT volumes. Original size and resolution of those CT volumes were $512, \times 512 \times (198 - 546)$ voxels and $(0.549 - 0.904) \times (0.549 - 0.904) \times 2\,\text{mm}^3$, respectively. We interpolated all CT volumes into $(281 - 463) \times (281 - 463) \times (396 - 762)$ voxels with 1 mm isotropic resolution.

The initial training rate was manually set as 1.0×10^{-3}, and 100,000 times of optimizations were performed with Adam optimizer. Patch sizes for 2D and 2D U-Nets are defined as 288×192 and $288 \times 192 \times 16$, respectively (See Fig. 5). We set the mini-batch size as 4. The parameters for the Watershed algorithm were manually set as $s = 0.01$ and $t = 0.2$.

We implemented the programs with Python 3.6.10 and TensorFlow 2.2.0. Those programs were executed with CUDA 10.1, cuDNN 7.6.5, NVIDIA Driver 465.19.01, and Ubuntu 18.04.5. A Quadro P6000 (NVIDIA) GPU is utilized for GPU-based processes. We used workstations with a Xeon E5-1680 v4 (Intel) CPU and 128 GB main memory (Sect. 3.2). Process parallelization between CT volumes was performed on the graphical process. Note that CPU-based inference and the subsequent graphical process are conducted only to measure the processing time. Their results are not used for other evaluations.

3.2 Evaluations

Segmentation Accuracies. The network estimates the distance maps (Sect. 2.2). Thresholding against the estimated distance maps using a low threshold $t (0 < t < 1)$ allows us to obtain intestine regions. We set $t = 0.01$ and quantitatively evaluate the segmentation accuracy. Since the ground-truth labels were traced only on several axial slices, evaluation is performed on those axial slices.

We evaluate the segmentation accuracy by the precision, the recall, and the Dice score. In addition, Wilcoxon signed-rank test is performed to compare each metric of those methods. The results are shown in Table 1.

System Usefulness. Our proposed method is utilized for diagnosis assistance of the ileus and the intestinal obstruction. It is essential to suggest diseased point candidates as segmentation results' endpoints, as illustrated in Fig. 1. In addition, the number of diseased point candidates should be small. Therefore, we defined two metrics, *the minimum distance (MD)* and *the number of diseased point candidates (NDPC).*

MD represents a distance between the diseased point and its nearest segmentation results' endpoints (colored red or blue). NDPC represents the number of diseased point candidates suggested to the clinician. The smaller MD and

Table 1. Segmentation accuracy metrics (precision, recall, and Dice) and their p-values.

Method	Precision	Recall	Dice
Proposed (2D U-Net)	0.971 ± 0.053	0.492 ± 0.065	0.650 ± 0.064
Previous (3D U-Net)	0.969 ± 0.060	0.497 ± 0.059	0.654 ± 0.059
(p-values)	0.582	0.035	0.105

Table 2. Metrics for system usefulness, minimum distance (MD) and number of diseased point candidates (NDPC) described in Sect. 3.2.

Method	MD	NDPC
Proposed (2D U-Net)	22.6 ± 17.3	17.3 ± 7.3
Previous (3D U-Net)	23.7 ± 17.2	17.4 ± 7.3
(p-values)	0.189	0.611

Table 3. Processing time (inference and graphical process) by GPU or CPU processing.

GPU/CPU	Method	Inference (minutes)	Graphical process (minutes)
GPU	Proposed (2D U-Net)	0.5 ± 0.1	23.2 ± 6.5
	Previous (3D U-Net)	4.2 ± 1.1	23.6 ± 6.5
	(p-values)	$<2.2 \times 10^{-16}$	0.0006
CPU	Proposed (2D U-Net)	2.5 ± 0.7	24.0 ± 6.6
	Previous (3D U-Net)	47.9 ± 12.9	25.5 ± 7.9
	(p-values)	$<2.2 \times 10^{-16}$	0.002

NDPC values represent the better results. A visual illustration of those metrics is included in Fig. 7. Those metrics' values are shown in Table 2. Wilcoxon signed-rank test is also performed.

Computational Time. We calculate the computational time for inference (Sect. 2.2) and the graphical process (Sect. 2.3). Wilcoxon signed-rank test is also performed to compare between the proposed (2D U-Net) and previous (3D U-Net) methods. Note that even if a GPU performed inference, the graphical process is conducted by a CPU.

As shown in Table 3, the proposed method's inference time was around eight times faster than the 3D U-Net's using a GPU, under the significant difference. The difference of inference times using a CPU was around 20-times. Although the mean graphical process times were similar for all conditions, significant differences were observed between GPU's and CPU's inference results.

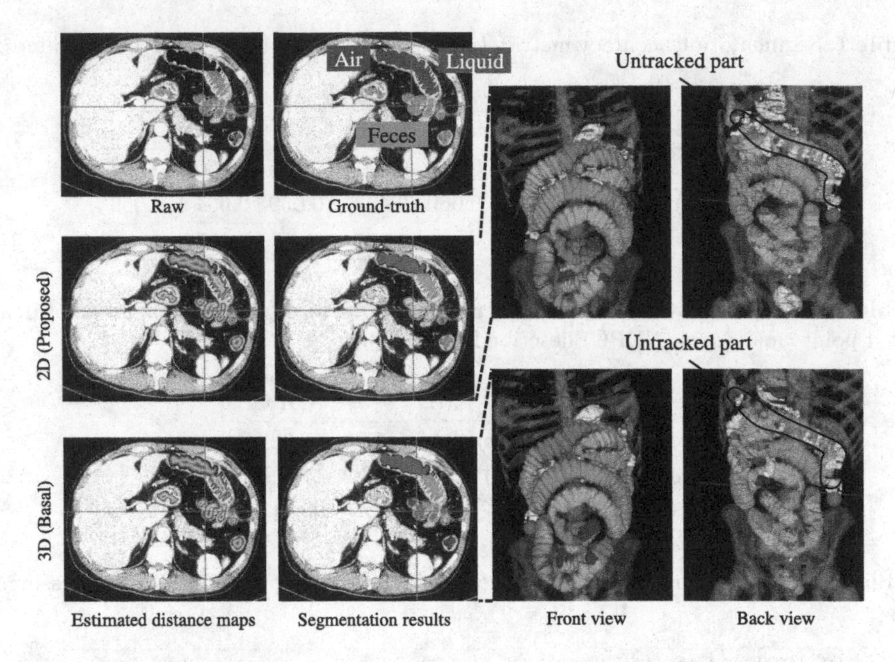

Fig. 8. Case that similar results were observed between proposed (2D U-Net) and previous (3D U-Net) methods. Both methods segmented and colored most regions correctly. Common issue was untracked part filled with feces. See Fig. 7 for legend.

Qualitative Analysis. Images and results of several CT volumes are visualized. Two examples are shown in Figs. 8 and 9.

4 Discussion

On the quantitative comparison, the average segmentation performances were very similar, as shown in Table 1. Furthermore, Table 2 suggests that those two methods' system usefulness is also comparable. Nevertheless, the proposed (2D U-Net) method's training and testing times were much shorter than the previous (3D U-Net) method, as shown in Table 3.

Figure 8 shows a case that similar results were observed between proposed (2D U-Net) and previous (3D U-Net) methods. Both methods segmented and colored the intestine's most parts correctly. A common issue was on an untracked part filled with feces. Those parts were segmented into many small regions. Since the graphical process (Sect. 2.3) analyze relations between segmentation results, those isolated segmentation results cannot be correctly analyzed. Those results imply that network selection does not affect segmentation results' connectivity.

Figure 9 shows a case that the proposed (2D U-Net)'s path estimation was incorrect. Since the proposed method's coloring was also incorrect, the segmentation results' part directing to the diseased point was not colored red or blue. This issue was due to the false connection between unconnected regions. Since

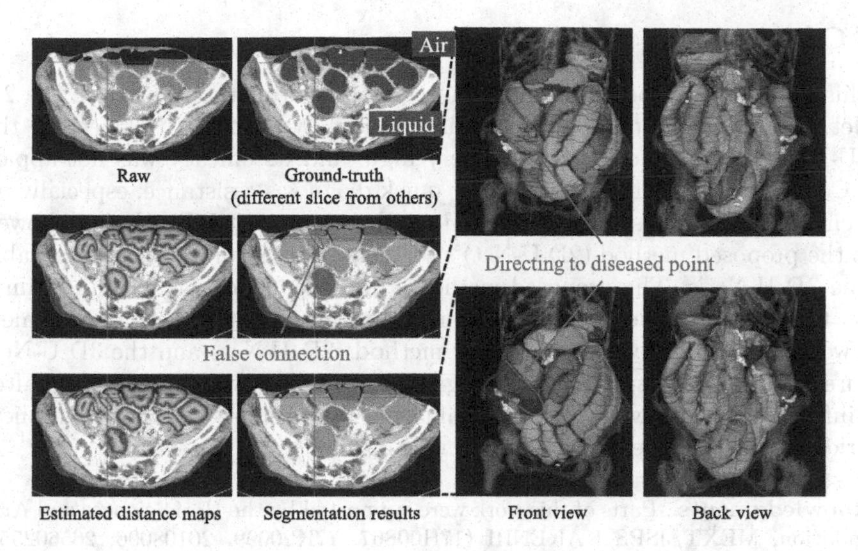

Fig. 9. Case that proposed (2D U-Net)'s path estimation was incorrect. Since ground-truth was sparsely generated, nearest slice to other axial slices is shown. Proposed method's coloring was wrong, and the segmentation results' part directing to diseased point was not colored in red or blue. This issue was due to false connection between unconnected regions, causing incorrect graphical process. See Fig. 7 for legend.

the connectivity between the regions was incorrect, the graphical process was also performed wrong. Although this was an example of the proposed method's error, Table 2 does not represent that the proposed method (2D U-Net)'s usefulness was significantly worse than the previous (3D U-Net)'s. Further quantitative evaluations of the graphical process are required in the future.

For the computational cost, Table 3 represents that the proposed method (2D U-Net)'s inference was much faster than the previous (3D U-Net) method on both GPU and CPU processing. As explained in Sect. 2.2, the 3D U-Net's number of parameters is more than 3-times than the 2D U-Net's. As a result, the 3D U-Net's computational cost was much larger than the 2D U-Net.

However, the graphical process requires more than 20 min on average for all conditions in Table 3. Therefore, speeding up the graphical process is crucial for practical use. Furthermore, it is also essential to investigate the acceptable time for practice use by taking questionnaires for many clinicians, especially for emergency diagnosis. Suppose the proposed method is utilized for clinical practice, inference and the graphical process are required to obtain a CT volume's segmentation results.

Totally, although some differences exist between the 2D and 3D U-Net's results, their differences are not significant (Tables 1 and 2). In contrast, reduction of the computational cost is considerable (Table 3). The results imply that the 2D U-Net is also promising for CPU processing.

5 Conclusions

In this paper, we proposed an intestine segmentation method based on the 2D U-Net. We alternated the 3D U-Net utilized in [5] by the 2D U-Net because the 3D U-Net's computational cost was very high, and its efficacy was not apparent. Computational time is essential for quick diagnosis assistance, especially on the emergency diagnosis. Experimental results using 110 CT volumes showed that the proposed method (2D U-Net)'s segmentation accuracy was comparable to the 3D U-Net's. The proposed method's inference time was around 8-times faster than the 3D U-Net's. Segmentation accuracies and other performance metrics were similar between the proposed method (2D U-Net) and the 3D U-Net. Future work includes speeding up the graphical process, which is required after the inference. Future work includes evaluating the proposed method using new metrics considering the seriousness or crisis of diseases.

Acknowledgments. Parts of this work were supported by the Hori Sciences and Arts Foundation, MEXT/JSPS KAKENHI (17H00867, 17K20099, 26108006, 26560255), JSPS Bilateral Joint Research Project, AMED (JP19lk1010036, JP20lk1010036), and JST CREST (JPMJCR20D5).

References

1. Shah, Z.K., Uppot, R.N., Wargo, J.A., Hahn, P.F., Sahani, D.V.: Small bowel obstruction: the value of coronal reformatted images from 16-multidetector computed tomography-a clinicoradiological perspective. J. Comput. Assist. Tomogr. **32**(1), 23–31 (2008)
2. Sainani, N.I., Näppi, J., Sahani, D.V., Yoshida, H.: Computer-aided detection of small bowel strictures for emergency radiology in CT enterography. In: Yoshida, H., Cai, W. (eds.) ABD-MICCAI 2010. LNCS, vol. 6668, pp. 91–97. Springer, Heidelberg (2011). https://doi.org/10.1007/978-3-642-25719-3_13
3. Zhang, W., Liu, J., Yao, J., Louie, A., Nguyen, T.B., Wank, S., Nowinski, W.L., Summers, R.M.: Mesenteric vasculature-guided small bowel segmentation on 3-D CT. IEEE Trans. Med. Imaging **32**(11), 2006–2021 (2013)
4. Shin, S.Y., Lee, S., Elton, D., Gulley, J.L., Summers, R.M., et al.: Deep small bowel segmentation with cylindrical topological constraints. In: Martel, A.L. (ed.) MICCAI 2020. LNCS, vol. 12264, pp. 207–215. Springer, Cham (2020). https://doi.org/10.1007/978-3-030-59719-1_21
5. Oda, H., et al.: Intestinal region reconstruction of ileus cases from 3D CT images based on graphical representation and its visualization. In: Medical Imaging 2021: Computer-Aided Diagnosis, vol. 11597, p. 115971N. International Society for Optics and Photonics (2021)
6. Çiçek, Ö., Abdulkadir, A., Lienkamp, S.S., Brox, T., Ronneberger, O.: 3D U-net: learning dense volumetric segmentation from sparse annotation. In: Ourselin, S., Joskowicz, L., Sabuncu, M.R., Unal, G., Wells, W. (eds.) MICCAI 2016. LNCS, vol. 9901, pp. 424–432. Springer, Cham (2016). https://doi.org/10.1007/978-3-319-46723-8_49
7. Ronneberger, O., Fischer, P., Brox, T.: U-net: convolutional networks for biomedical image segmentation. In: Navab, N., Hornegger, J., Wells, W.M., Frangi, A.F. (eds.) MICCAI 2015. LNCS, vol. 9351, pp. 234–241. Springer, Cham (2015). https://doi.org/10.1007/978-3-319-24574-4_28

Generation of Patient-Specific, Ligamentoskeletal, Finite Element Meshes for Scoliosis Correction Planning

Austin Tapp[1(✉)], Christian Payer[2], Jérôme Schmid[3], Michael Polanco[1], Isaac Kumi[1], Sebastian Bawab[1], Stacie Ringleb[1], Carl St. Remy[4], James Bennett[4], Rumit Singh Kakar[5], and Michel Audette[1]

[1] Old Dominion University, Norfolk, VA 23529, USA
atapp001@odu.edu
[2] Graz University of Technology, Inffeldgasse 16, 8010 Graz, Austria
[3] University of Applied Sciences and Arts of Western Switzerland, Avenue de Champel 47, 1206 Geneva, Switzerland
[4] Children's Hospital of the King's Daughter's, 601 Children's LN, Norfolk, VA 23529, USA
[5] Oakland University, Rochester, MI 48309, USA

Abstract. Finite element (FE) biomechanical studies for adolescent idiopathic scoliosis (AIS) treatments will greatly benefit from utilizing true-scale, patient-specific anatomy that accurately characterizes all tissue properties. This study presents a method to automatically generate patient-specific, FE meshes containing volumetric soft tissues, such as ligaments and cartilage, that are inconspicuous in computed tomography (CT) imaging of AIS patients. Convolutional neural network (CNN) derived vertebrae segmentations, obtained from CT scans, provide a foundation for subsequent elastic deformations of ligamentoskeletal, computer-aided designed (CAD) surface meshes, to ascertain patient-specific anatomy, including soft tissue structures. Patient-specific, ligamentoskeletal meshes are then tetrahedralized for use in FE studies. Dice similarity coefficients of 90% and submillimeter Hausdorff distances demonstrate vertebrae and intervertebral disc fitting accuracy of the automatic methodology. In severe AIS cases, when CNN segmentations fail due to overfitting, a semi-automatic step augments the automatic method. The generated FE meshes can provide the basis for biomechanical simulations seeking to correct AIS through bracing, minimally invasive operations, or patient-specific, surgical procedures, like posterior spinal fusion.

Keywords: Adolescent idiopathic scoliosis · Patient-specific · Ligamentous meshing · Finite elements · Pre-surgical planning · Enhanced CT visualization

1 Introduction

Adolescent idiopathic scoliosis (AIS) affects 2 to 4% of populations aged 10 to 18 [4]. Individuals with AIS have "S" or "C"-shaped spinal curvatures, uneven shoulders, and offset hips. If AIS is not resolved, further complications may result in severe back

C. Oyarzun Laura et al. (Eds.): CLIP/DCL/LL-COVID/PPML 2021, LNCS 12969, pp. 13–23, 2021.
https://doi.org/10.1007/978-3-030-90874-4_2

problems as well as lung and heart damage. AIS treatments largely depend on patient's primary, lateral spine curvature, but skeletal maturity is also considered [13]. In mild cases, defined by curvature of 10 to 25°, physical therapy is often sufficient [13].

For growing individuals with moderate scoliosis, having spinal curvatures between 25 and 45°, brace treatment is required to rectify deformities and prevent progression [13]. Finite element (FE) modeling is frequently used to personalize and optimize bracing treatment; however, FE studies understate anatomical complexity [16]. It is known that the soft tissues surrounding the spine play an important role in governing correction, but soft tissue volumes in FE studies are represented as crude springs, one-dimensional rods, or simplified elastic shell elements, all of which are set with onerous, hand-designated anchor points [11, 15]. Further, FE bracing studies are based on patient-specific models developed by manual methods [6]. To improve and streamline individualized bracing treatments guided by patient-specific FE studies, the FE meshes utilized should be automatically produced and replace simplistic ligament representations with true representations of relevant anatomy, namely volumetric ligaments [16].

Severe AIS deformities, determined by curvatures greater than 45 degrees, must be resolved with surgical intervention [13]. For AIS patients with immature bone status, more treatment options are available [4]. However, most AIS must be treated with invasive procedures such as anterior or posterior spinal fusion (PSF) surgery [4, 13]. Surgical outcomes seek to prevent progression, maintain coronal and sagittal alignment, level the shoulders, correct the spinal deformity, and preserve motion segments. Computed tomography (CT) or biplanar X-ray imaging is deemed appropriate to evaluate curvature and determine a correction strategy [8]. Unfortunately, such images provide limited insight into how patients will respond to state-of-the-art correction procedures like PSF, shown in Fig. 1 [11]. Thus, extra steps to mobilize the spine are performed intra-operatively, increasing patient morbidity, operating room time, and blood loss [12]. Similar to bracing studies, FE simulations predicting patient responses to corrective surgery can make operations safer and more efficient [10]. However, biomechanical simulations cannot offer corrective force estimations without true, multi-material, volumetric meshing [2]. Patient-specific FE studies must include realistic soft tissues for accurate pre-surgical planning and faithful predictions of corrective outcomes [11].

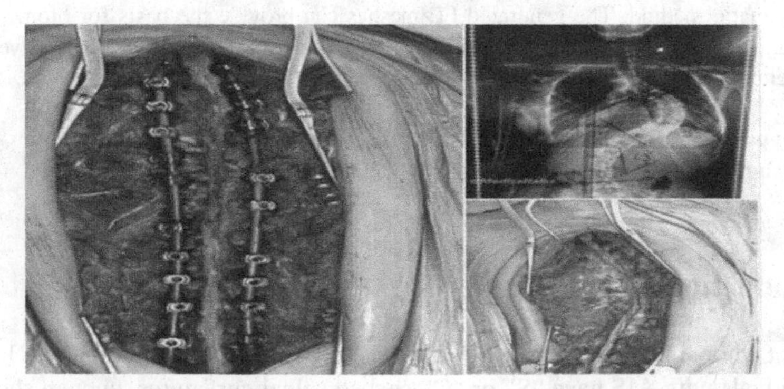

Fig. 1. (Bottom right) Before and (left) after views of a posterior spinal fusion operation. This procedure corrects severe scoliosis, which is diagnosed, in part, with X-ray images (top right).

Whatever their intended use, FE meshes designed for scoliosis correction studies must appropriately consider anatomical elements, including soft tissues, as volumes. While volumetric segmentation of vertebral bodies and intervertebral discs has been demonstrated, even in scoliotic cases, studies producing ligamentoskeletal meshes from asymptomatic or symptomatic cases are not found [7, 9]. Therefore, this study proposes a methodology for the generation of patient-specific, ligamentoskeletal FE meshes that provide true and realistic foundations for biomechanical simulations. The FE meshes are generated from CT scans, which are routinely obtained by clinicians attempting AIS correction. The presented approach is automatic for asymptomatic to moderately scoliotic cases, obviating hand-designation of soft tissues in FE meshes used by studies considering ideal bracing treatments. Application of the method to severe cases currently necessitates the addition of one semi-automatic step. This minor addition greatly bolsters the segmentation abilities of a pre-trained neural network (NN), which constitutes the basis of the method. Pre-operative planning for severe AIS can be greatly enhanced by using biomechanical simulations harnessing ligamentoskeletal FE meshes. Pre-operative planning using ligamentoskeletal meshes is absent from literature, but planning using simplified FE meshes is not [16]. Presumably, simulations with comprehensive FE meshes should better define ideal treatments, provide more specific procedural guidelines, and could elucidate the use of minimally invasive correction techniques on AIS patients with mature bone status, but without detailed patient anatomy, exploring therapy options in a risk-free manner, *in silico*, is nearly impossible [10, 11].

2 Methods

2.1 Overview

The presented approach requires a CT scan to generate FE, ligamentoskeletal meshes. An anatomist-drawn, computer-aided design (CAD) template mesh is initially positioned within the CT's image space based on NN-derived vertebrae segmentations. The CAD mesh is then deformed using CT information, such as image gradients and voxel intensities, producing a patient-specific, ligamentoskeletal mesh. The patient-specific meshes are tetrahedralized into FE meshes. This overview is illustrated in Fig. 2.

2.2 Patient-Specific, Ligamentoskeletal, Finite Element Mesh Generation

Computer-Aided Design Template. Due to the inconspicuous presentation of soft tissues in CT imaging, segmenting ligaments by intrinsic voxel-based or NN techniques is precluded. Therefore, a top-down, model-based segmentation approach is required. The approach begins with an anatomist-developed CAD model of a human torso, which was constructed with careful consideration of cadavers and multi-modal torso images, while using ratified anatomy and physiology textbooks (cghero.com). The torso is consistent in structure with a healthy, average, adult human to permit generalized application of the current methodology to a wide range of cases. The CAD torso contains all bone, ligament, cartilage, and other anatomy as accurate and realistic meshes. While the patient-specific, mesh generation process scales and elastically deforms the torso mesh, theoretically allowing its use on any patient's case, an adolescent torso mesh may be more appropriate for clinical use with AIS individuals. However, this study uses an adult torso since most CT images were obtained from adult patients. Before the CAD meshes are deformed, vertebral segmentations are needed to correctly position the meshes in CT image space. Segmentations are attained either automatically, through a NN, or with a minimally supervised step that corrects NN overfitting, which occurs when severe AIS cases are tested.

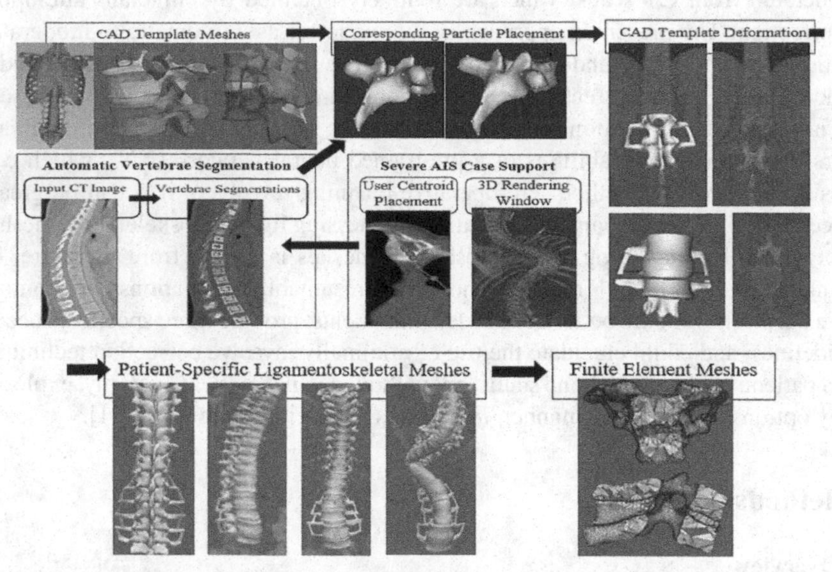

Fig. 2. Overview of the presented method. From left to right (top) a view of the full CAD torso, a thoracic and a lumbar spine segment, and (bottom) the neural network used for automatic segmentations. In severe AIS cases, a semi-automatic addition provides vertebrae centroids to the NN. Then, both CAD meshes and vertebrae segments are characterized with corresponding particles that guide an affine transform; this is followed by elastic deformations. After CAD template deformation, combined patient-specific, ligamentoskeletal meshes with volumetric soft tissues are subsequently tetrahedralized into FE meshes.

Automatic Vertebrae Segmentation. Vertebral segmentations from CT images are obtained with a fully automatic, 3-step, coarse-to-fine NN, whose architecture is shown in Fig. 3. The first step predicts approximate x and y coordinates of the spine as $\hat{x}_{spine} \in \mathbb{R}^2$ using a U-Net variant that has average, instead of max pooling, and linear upsampling, instead of transposed convolutions. The variant consists of five levels; convolutional layers have kernel sizes of [3 × 3 × 3], 64 filter outputs, and zero padding, for equal input and output sizes. The U-Net performs heatmap regression on a single target volume $\dot{h}_{spine}(x; \sigma_{spine})$: $\mathbb{R}^3 \rightarrow \mathbb{R}$, which is formed by merging Gaussian heatmaps with size σ_{spine} of all individual vertebrae target coordinates $\dot{x}_i$. L2-loss minimizes the difference between the target heatmap volume $\dot{h}_{spine}$ and a predicted heatmap volume $\hat{h}_{spine}(x)$: $\mathbb{R}^3 \rightarrow \mathbb{R}$. The final predicted spine coordinate $\hat{x}_{spine}$ is the x and y coordinate of the center of mass of $\hat{h}_{spine}$ [14].

The second step localizes the center of the vertebral bodies through the SpatialConfiguration-Net (SC-Net), which is comprised of 2 components that work together to determine local landmark appearance while considering the landmark's spatial configuration [14]. The component determining local landmark appearance is made of 5 levels, 2 convolutional layers followed by a downsampling and two more convolutional layers. Convolutional layers use leaky ReLU activation functions, kernel sizes of [3 × 3 × 3] and 64 filter outputs. The spatial configuration component has 4 convolutions with [7 × 7 × 7] kernels. The entire SC-Net completes heatmap regression on N target vertebrae v_i ($i = 1 \ldots N$) so each target coordinate $\dot{x}_i$ is represented as a Gaussian heatmap volume $\dot{h}_i(x; \sigma_i)$: $\mathbb{R}^3 \rightarrow \mathbb{R}$, centered at $\dot{x}_i$. Simultaneously, SC-Net predicts all N output heatmap volumes $\hat{h}_i(x)$: $\mathbb{R}^3 \rightarrow \mathbb{R}$. A modified L2-loss function permits learning of individual σ_i values for the target volumes $\dot{h}_i$. The predicted spine coordinate $\hat{x}_{spine}$ of the first step is used to narrow the portions of the CT processed. Heatmap predictions are thresholded for local maxima, outputting a final centroid. Superior and inferior vertebral centroids are determined by heatmaps closest to the top and bottom of the volume, respectively. The overall centroid sequence is output after predicted centroids are merged and filtered based on proximity criteria that removes unlikely predictions.

Finally, individual vertebrae are segmented with the U-Net variant described above. The centroids output by SC-Net provide a semantic label that crops the region around the landmark, allowing vertebrae to be segmented at full resolution and independently of one another. The resulting output binary segmentations are transformed and resampled to their original input position in the overall CT volume.

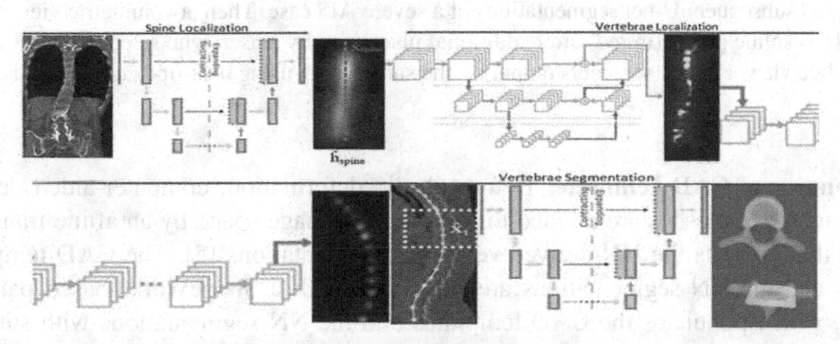

Fig. 3. Architecture of the NN that automatically segments vertebrae. The coarse-to-fine network begins with spine localization (top left) and ends with vertebrae segmentation (bottom right).

Support for Severe AIS Segmentations. When testing severe AIS cases, like in Fig. 4, the NN can suffer from quasi-overfitting. The NN lacks generalization for severe AIS cases because such training data is extremely uncommon. Therefore in patients with severe curvatures, vertebrae may be quite close to one another, resulting in incorrect filtering, poor centroid placement, and subsequent segmentation failure. To address this issue, a clinician-friendly semi-automatic algorithm was developed. The algorithm imports all centroids and displays them in a graphical user interface. Initially, centroids act as points defining a parametric spline. Users provide the spline S: [a, b] $\rightarrow \mathbb{R}^3$ with additional k ordered subinterval control points $[t_i, t_{i+1}]$, i = 0,…, k − 1 such that the interval, [a, b] = $[t_0, t_1] \cup [t_1, t_2] \cup [t_{k-2}, t_{k-1}] \cup [t_{k-1}, t_k]$, defines all polynomials comprising the spline S. Alternatively, users can discard control points of the spline. While users visualize the spline, centroids, and volume-rendered CT in 3-dimensions, user interaction occurs inside an image slice window, which scrolls axially through the CT and allows for controlled centroid placement and removal. Removal is straightforward; a new spline is defined by remaining control points. For placement, the point along the spline that is closest to the user-defined point determines a new control point. In other words, when users select a position in the slice view, the algorithm uses the coordinates of that position to calculate the nearest 3D coordinate that appears on the spline. Then, user-defined points are adjusted to that nearest 3D coordinate, resulting in shape-aware centroid positioning that considers the curvature of a patient's spinal column. Centroids are passed to the NN's U-Net, enhancing its segmentation abilities in severe AIS cases.

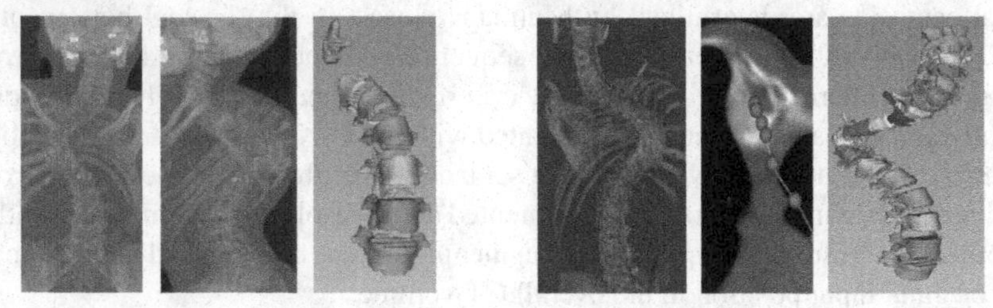

Fig. 4. (Left to right) Two views of centroids placed automatically by the vertebrae localizing SC-Net and subsequent U-net segmentations of a severe AIS case. Then, a volumetric view of the centroids as spline points (in red) after additional placement by a user, who selected points using the 2D slice view. Finally, segments output by the same U-net using user-updated centroids.

Deformation of CAD Template. Before elastic deformation, computer-aided design (CAD) template meshes are aligned to relevant CT image space by an affine transformation that exploits the NN-derived vertebrae segmentations [5]. The CAD template meshes and the NN segmentations are processed with a ShapeWorks-based particle system, which populates the CAD templates and the NN segmentations with surface particles that spatially correspond between like structures [3]. The set of corresponding particles are geometrically positioned by minimizing a cost function, which combines ensemble shape correspondence and consistent surface sampling, defined as $Q = \alpha H(Z) - \sum^{N}_{n=1} H(X_n)$. For more details, see Cates et al. [3]. When positioned, particles fully encapsulate shapes' surface geometry. Corresponding particles between the CAD mesh and NN-derived segmentations are homologous points that guide the affine transform.

After the CAD mesh is affinely aligned onto relevant CT image space, elastic deformation of the CAD mesh occurs. The CAD mesh is deformed using a framework of physically-based triangular meshes. All CAD mesh vertices are lumped as a single mass component and are driven toward pertinent CT voxels. The meshes' vertex motions are solved using an implicit Euler scheme that determines Newtonian dynamics-based forces described by $f_i = \alpha_i(R_i - P_i)$, where P_i is a vertex position and f_i is the force that attracts

P_i towards its target vertex, R_i. For more details about how a vertex's state, position and motion is determined, the reader is directed to Damopoulos et al. [5]. Weighting factors considering strong image gradients and upper-quartile voxel intensities encourage deformation of the vertebrae portion of the CAD toward highly conspicuous aspects of the CT image, i.e., the vertebrae of the CT. Simultaneously, anatomy surrounding the vertebral portions of the CAD model is locally deformed due to the vertices mass grouping, resulting in context-aware, soft tissue positioning. This surmises positions of the soft tissues, such as ligaments, that are inconspicuous in the CT image. Additionally, the deformation process applies regularization to penalize inappropriate mesh distortions, incorporates internal forces to promote smoothing and shape similarity, and models local vertex geometries with mean value encoding.

Conversion of Patient-Specific Mesh into Finite Elements. Following elastic deformation, the ligamentoskeletal CAD mesh is now patient-specific. Each anatomical structure of the CAD mesh may be split from its associated vertebral parent and tetrahedralized, or a complete mesh can be tetrahedralized. The volumetric conversion is performed using a modified form of CGAL's 3D variational mesh generation script [1]. While 1D features, like edges and surface boundaries, are preserved, users may denote facet and tetrahedra sizes, allowing FE meshes to be customized. For this study, parameters set were an edge size of 0.025, facet angle of 30, facet size of 100, facet distance of 0.05, cell radius edge ratio of 6, and a cell size of 1000.

3 Results

3.1 Datasets

NN training used the CT dataset of the VerSe 2020 grand challenge (verse2020.grand-challenge.org/). Training and testing was performed with Tensorflow (tensorflow.org/) using a mini-batch size of 1 for all networks, 10,000 iterations for the spine localization network, 50,000 iterations for the vertebrae localization network and 50,000 iterations for the vertebrae segmentation network. Modified U-Nets used the Adam optimizer at a 10^{-4} learning rate, while SC-Net used the Nesterov optimizer at a 10^{-8} learning rate.

A randomized sample of asymptomatic CT scans, obtained from a mixture of SpineWeb and VerSe challenge datasets, was used to evaluate deformation accuracy of patient-specific, CAD template vertebrae (spineweb.digitalimaginggroup.ca/). Additionally, a cohort of 5 mild/moderately scoliotic CTs, found in the VerSe dataset, was used to evaluate the deformation accuracy of patient-specific, CAD template vertebrae for symptomatic cases. In total, 93 deformed CAD vertebrae, generated from existing asymptomatic or symptomatic CTs, were evaluated with publicly available ground truth vertebral segmentations. In some cases, CT images covered the entire spinal column, while other images only included portions. Overall, regions of the cervical spine, thoracic spine and lumbar spine were fit 2, 6, and 10 times, respectively.

For soft tissue evaluation, cross-modality, CT and magnetic resonance imaging (MRI) data from SpineWeb was included as a part of the random sample data mentioned above. When deforming a CAD template to a CT image with an associated MRI, the intervertebral discs (IVDs) from the patient-specific, ligamentoskeletal CAD mesh were compared to anatomist-provided IVD ground truths, segmented from the MRIs. The cross-modality dataset images were limited to the lumbar region of the spine, from thoracic vertebra 12 to the sacrum and were used to evaluate 20 IVDs. Evaluation of the other soft tissue structures in the ligamentoskeletal meshes cannot yet be completed. For now, patient-specific, ligamentoskeletal meshes can only be evaluated on anatomy, like IVDs, that are conspicuous in available imaging. To authenticate remaining soft tissue, a validation strategy that makes use of cadavers is planned.

To quantitatively evaluate the presented method during the testing of severe AIS cases, synthetic severe AIS cases were created through a novel VTK-based spine deformation program (vtk.org). The synthetic AIS cases included Lenke classifications of 3A, 1B, 1C, 2C, 3C, 4C, 5 and 6 and had associated vertebrae ground truths. These synthetic spines provided 136 vertebrae ground truths for severe AIS case validation.

3.2 Quantitative Results

All obtained, patient-specific, ligamentoskeletal meshes were evaluated for fitting accuracy with two metrics, Dice similarity coefficient (DSC) and average Hausdorff distance (aHD). These metrics compare a deformed mesh vertebra or IVD to its respective ground truth segmentations in the patient CT. Prior to this comparison, the meshes are converted into a segmentation. DSC volumetrically evaluates the number of segmentation elements that overlap to the total elements found in the segmentations; the DSC of a segment compared to itself would be 1. aHD evaluates the overall surfaces of both segmentations, measuring disparity, in millimeters, between surfaces of a CAD mesh segmentation and surfaces of its corresponding ground truth. DSC of vertebrae generated from asymptomatic data achieved an average score of 0.84 and aHD of 1.42 mm. For symptomatic cohorts, DSC and aHD were slightly lower, averaging 0.77 and 1.54 mm. IVDs generated with CT imaging averaged 0.88 and 1.12 mm, for DSC and aHD, respectively. Figure 5 displays some outcomes, while Table 1 summarizes all results.

The presented method typically excels in HD and is competitive in DSC score when correctly compared to related studies [7, 9]. Korez et al. used deformable models to produce vertebrae segmentations from asymptomatic cases, achieving an average DSC of 0.93 and aHD of 3.83 mm [9]. Guerroumi et al. did not use deformable models, but performed direct NN segmentation on scoliotic cases, reporting DSC of 0.82 and 0.74 for vertebrae and IVDs, respectively; HD was not reported [7]. Notably, these studies evaluated just vertebral bodies (VBs), whereas the present study evaluated vertebrae in their entirety (All). For direct comparison, VBs obtained by the presented method were evaluated. In asymptomatic and symptomatic cases, VBs achieved an average DSC of 0.90 and 0.82 and an aHD of 0.86 and 1.31 mm. Most likely, improved HD are due to the presented methods implementation, which emphasizes vertebral shape. The compared methods “grew” segmentations, by increasing models to match predetermined volumes (Korez et al.) or directly segmented anatomy from 2D images (Guerroumi et al.) [7, 9].

Table 1. The quantitative overview. Top two columns show results from related studies [7, 9]. Bottom three columns show presented results. A **Yes** indicates studies used scoliotic cohorts (Scoliotic?) or deformable model registration (Deformable?). DSC and aHD Evaluations were performed on vertebral bodies (VBs), all vertebral anatomy (All) or intervertebral discs (IVDs).

Method	Scoliotic?	Deformable?	DSC (out of 1)	aHD (mm)
[7]	**Yes**	No	VBs: 0.82 IVDs: 0.74	Not Tested
[9]	No	**Yes**	VBs: 0.93	VBs: 3.83
Presented	No	**Yes**	VBs: 0.9 All: 0.84	VBs: 0.86 All: 1.42
Presented	No	**Yes**	IVDs: 0.88	IVDs: 1.12
Presented	**Yes**	**Yes**	VBs: 0.82 All: 0.77	VBs: 1.31 All: 1.54

3.3 Qualitative Results

Symptomatic cases lacking ground truths found in the VerSe challenge dataset are shown in Fig. 6. The images demonstrate further applications of this method. Considering quantitative results and the appearance of the meshes shown, it is likely the models have been generated correctly, but validation remains to be completed. Figure 6 also validates FE software compatibility of FE meshes by FEBio (febio.org).

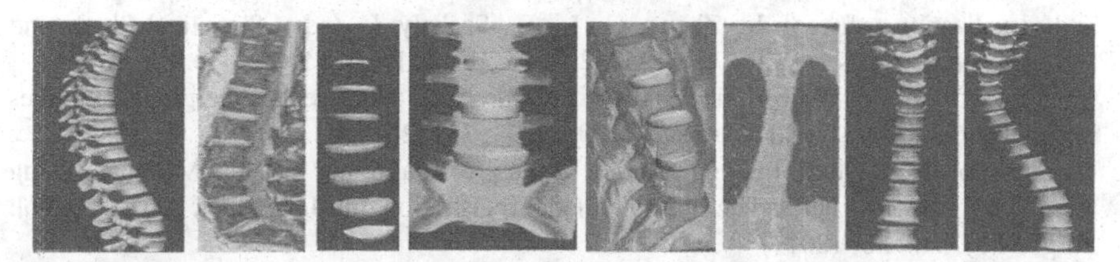

Fig. 5. (Left to right) Screenshot of vertebrae fitting on an asymptomatic CT scan, IVDs fit from a CAD to CT and MRI ground truth. A volumetric CT with 3 IVDs and, to the right, its companion MRI with the same CAD IVDs. Then, vertebrae fit onto to a true symptomatic CT case and (far right) a synthetic case. Ground truth is red, background is blue, meshes are white. (Color figure online)

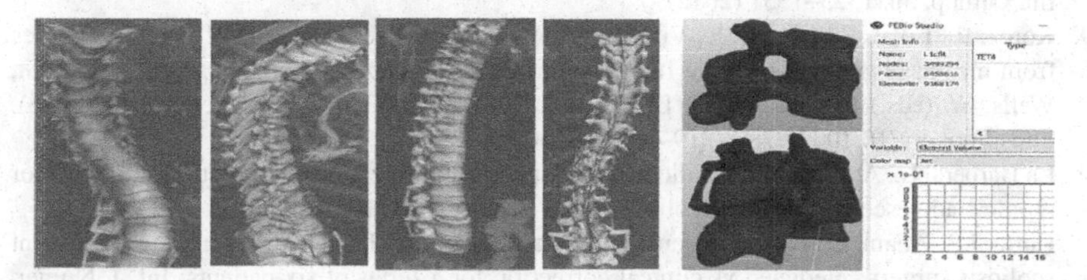

Fig. 6. Patient-specific, ligamentoskeletal meshes. Far right, FE analyzed by FEBio's mesh inspector tool that shows meshes are TET4 and blue, which reflects a low minimum edge ratio. (Color figure online)

4 Conclusion

This study presents a methodology for the generation of ligamentoskeletal, patient-specific, multi-surface FE meshes that contain true-scale, soft tissue anatomy. For moderate AIS cases, these methods provide an automatic, robust foundation to brace treatment modeling. For severe cases, these methods offer a semi-automatic design that will enhance pre-surgical planning and may improve outcomes. In either case, FE biomechanical simulations using true representations of soft-tissue volumes that are provided by this method will better predict patient-specific responses to corrective treatment. Further, this methodology is relatively generic and adaptable; it may provide additional advancements for the orthopedic presurgical planning community, especially for predictive studies in other ligamentous areas, such as the shoulder and knee.

References

1. Alliez, P., Cohen-Steiner, D., Yvinec, M., Desbrun, M.: Chapter 10: Variational tetrahedral meshing. In: ACM SIGGRAPH 2005 Courses, SIGGRAPH 2005 (2005)
2. Audette, M.A., et al.: Towards a deformable multi-surface approach to ligamentous spine models for predictive simulation-based scoliosis surgery planning. In: Zheng, G., Belavy, D., Cai, Y., Li, S. (eds.) CSI 2018. LNCS, vol. 11397, pp. 90–102. Springer, Cham (2019). https://doi.org/10.1007/978-3-030-13736-6_8
3. Cates, J., Elhabian, S., Whitaker, R.: ShapeWorks: particle-based shape correspondence and visualization software. In: Zheng, G., Li, S., Szekely, G. (eds.) Statistical Shape and Deformation Analysis, vol. 2017, pp. 257–298. Academic Press, MA (2017). Ch. 10
4. Cheung, Z. B., Cho, S.: Idiopathic scoliosis in children and adolescents: emerging techniques in surgical treatment. World Neurosurg. **130**, e737–e742 (2019)
5. Damopoulos, D., Schmid, J.: Segmentation of the proximal femur in radial MR scans using a random forest classifier and deformable model registration. Int. J. Comput. Assist. Radiol. Surg. **14**(3), 545–561 (2019)
6. Guan, T., Zhang, Y., et al.: Determination of three-dimensional corrective force in adolescent idiopathic scoliosis and biomechanical finite element analysis. Front. Bioeng. Biotechnol. **8**, 963 (2020)
7. Guerroumi, N., Playout, C., et al.: Automatic segmentation of the scoliotic spine from MR images. In: International Symposium on Biomedical Imaging, vol. 2019, pp. 480–484 (2019)
8. Illés, T., Somoskeöy, S.: The EOS^{TM} imaging system and its uses in daily orthopaedic practice. Int. Orthop. **36**, 1325–1331 (2012)
9. Korez, R., Likar, B., Pernuš, F., Vrtovec, T.: Model-based segmentation of vertebral bodies from mr images with 3D CNNs. In: Ourselin, S., Joskowicz, L., Sabuncu, M.R., Unal, G., Wells, W. (eds.) MICCAI 2016. LNCS, vol. 9901, pp. 433–441. Springer, Cham (2016). https://doi.org/10.1007/978-3-319-46723-8_50
10. La Barbera, L., Aubin, C.E.: In silico patient-specific optimization of correction strategies for thoracic adolescent idiopathic scoliosis. Clin. Biomech. **81**, 105200 (2021)
11. Little, J.P., Adam, C.: Patient-specific computational biomechanics for simulating adolescent scoliosis surgery: predicted vs clinical correction for a series of six patients. Int. J. Numer. Methods Biomed. Eng. **27**(3), 347–356 (2011)
12. Lenke, L.G., Blanke, K.: Adolescent idiopathic scoliosis. A new classification to determine extent of spinal arthrodesis. J. Bone Joint Surg. **83**(8), 1169–1181 (2001)

13. Mohamed, M., Trivedi, J., Davidson, N., Munigangaiah, S.: Adolescent idiopathic scoliosis: a review of current concepts. Orthop. Trauma **34**(6), 338–345 (2020)
14. Payer, C., Urschler, M.: Integrating spatial configuration into heatmap regression based CNNs for landmark localization. Med. Image Anal. **54**, 207–219 (2019)
15. Rajaee, M.A., Arjmand, N., Shirazi-Adl, A.: A novel coupled musculoskeletal FE model of the spine – critical evaluation of trunk models. J. Biomech. **119**, 110331 (2021)
16. Wang, W., Baran, G.R., Cahill, P.J.: The Use of finite element models to assist understanding and treatment for scoliosis: a review paper. Spine Deformity **2**(1), 10–27 (2014)

Bayesian Graph Neural Networks for EEG-Based Emotion Recognition

Jianhui Chen, Hui Qian, and Xiaoliang Gong(✉)

Tongji University, Shanghai 201804, China
gxllshsh@tongji.edu.cn

Abstract. Emotion recognition has great significance in human-computer interaction, affective computing and clinical medicine, etc. Electroencephalography (EEG) is the most important one for emotion recognition due to its high temporal resolution. The progress in geometric deep learning provide powerful tool to explore the spatial features between EEG channels. There have been some studies using Graph-based methods, but neither do they reveal the latent structure of brain regions nor they contain uncertainty information. In this paper, we proposed a Bayesian Graph Neural Networks framework combined with a Sparse Graph Variational Auto-encoder. Our model can detect the latent communities between EEG channels in a non-parametric Bayesian way and provide uncertainty information of model prediction. Extensive experiments have been conducted to justify the effectiveness of our model and the results show that uncertainty information can help a lot.

1 Introduction

Emotion are important in the daily life of human as they play an important role in decision-making, perception, human interaction, etc. [1]. Emotion recognition has become an active field for its significance in human-computer interaction, affective computing, clinical medicine, etc. A vast number of studies have been conducted on EEG based emotion recognition [2–5]. However, the topological structures of EEG channels are not effectively exploited to learn more discriminative EEG representations in most existing EEG-based emotion recognition approaches.

Graph Neural Networks (GNN) which can effectively process unregular data that Convolutional Neural Networks (CNN) and Recurrent Neural Networks (RNN) cannot have gained tremendous interests [6–10]. GNN has been applied in [11] to capture inter-channel relations using an adjacency matrix learned by neural network. After that, more and more attentions are paid to graph theory-based methods. Zhong et al. [12] propose a regularized graph neural network (RGNN) for EEG-based emotion recognition with two regularizers, namely node-wise domain adversarial training (NodeDAT) and emotion-aware distribution learning (EmotionDL), to better handle cross-subject EEG variations and noisy labels. Ding et al. [13] propose LGG, a neurologically inspired graph neural network, to learn local-global-graph representations from Electroencephalography (EEG) for a Brain-Computer Interface (BCI). Wang et al. [14] used phase-locking

C. Oyarzun Laura et al. (Eds.): CLIP/DCL/LL-COVID/PPML 2021, LNCS 12969, pp. 24–33, 2021.
https://doi.org/10.1007/978-3-030-90874-4_3

value (PLV) graph convolutional neural networks (P-GCNN) to perform EEG emotion classification.

These works have achieved great improvement in classification accuracy, but they are less helpful in understanding the mechanism of neural activities during emotion processing, nor they can provide the uncertainty of classification results, which may prone to over-fitting due to limited data. In this paper, we proposed a Sparse Graph Variational Auto-encoder (SGVAE) based Bayesian Graph Neural Networks (BGNN) to find the underlying structure of different brain regions as well as perform emotion classification with uncertainty information.

2 Methods

In this section, we first propose the BGNN model and SGVAE model then we present the training and predicting algorithms for BGNN.

2.1 Bayesian Graph Neural Networks

Bayesian neural networks (BNNs) aim to capture model uncertainty of Deep neural networks (DNNs) by placing prior distributions over the model parameters to enable posterior updates during DNN training. It has been shown that these Bayesian extensions of traditional DNNs can be robust to over-fitting and provide appropriate prediction uncertainty estimation [15] which is especially effective for EEG dataset which does not have enough data and have strong individual variations.

Bayesian Graph Convolutional Neural Networks which is a Bayesian version of GNNs are first proposed in [16] for semi-supervised nodes classification. For emotion recognition, our goal is to compute the probability of labels Y^* given the training data X, Y together with the graph $\mathcal{G}_{obs}$ (which will be introduced later) and the test data X^* without labels:

$$p(Y^*|X^*, X, Y, \mathcal{G}_{obs}) = \int p(Y^*|X^*, W, \mathcal{G})p(W|X, Y, \mathcal{G})p(\mathcal{G}|\mathcal{G}_{obs})d\mathcal{G}dW$$
$$= \int p(Y^*|Z^*, W)p(W|Z, Y)f(Z|X, \mathcal{G}, \theta)f(Z^*|X^*, \mathcal{G}, \theta)p(\mathcal{G}|\mathcal{G}_{obs})d\mathcal{G}dW \quad (1)$$

where $p(\mathcal{G}|\mathcal{G}_{obs})$ is a parametric graph generative model from which we can sample graph $\mathcal{G} \sim p(\mathcal{G}|\mathcal{G}_{obs})$. $f(Z|X, \mathcal{G}, \theta)$ is a deterministic function extracting topological features Z and can be modeled by GNNs. The term $p(Y^*|Z^*, W)$ can be modelled by (Multi-layers perceptron) MLP with its parameters denoted as W. $p(W|Z, Y)$ is the posterior distribution of networks parameters. The Eq. (1) has shown that we only need to perform posterior inference on the parameters after GNN given the extracted features Z. While in [16], they perform posterior inference on the parameters of entire model which is inefficient. The integral in Eq. (1) is intractable so we approximate it by Monte Carlo

$$p(Y^*|X^*, X, Y, \mathcal{G}_{obs}) \approx \frac{1}{N_G S} \sum_{i=1}^{N_G} \sum_{j=1}^{S} p(Y^*{}_{i,j}|Z^*{}_{i,j}, W_{i,j})f(Z^*{}_{i,j}|X^*, \mathcal{G}_i, \theta) \quad (2)$$

where $W_{i,j}$ are sampled from $p(W|Z, Y)$ and $\mathcal{G}_i$ is sampled from $p(\mathcal{G}|\mathcal{G}_{obs})$.

Unfortunately, the posterior inference of $p(W|Z, Y)$ is also intractable. To perform posterior inference efficiently, Y. Gal et al. [15] and A. Hasanzadeh et al. [17] proposed that dropout and other stochastic regularization techniques are equal to some kind of approximate variational posterior distribution as long as we choose the appropriate prior. We adopt the concrete dropout method [21] to compute $p(W|Z, Y)$. The details of parametric graph generative model $p(\mathcal{G}|\mathcal{G}_{obs})$ will be introduced in next section.

The network architecture is shown in Fig. 1.

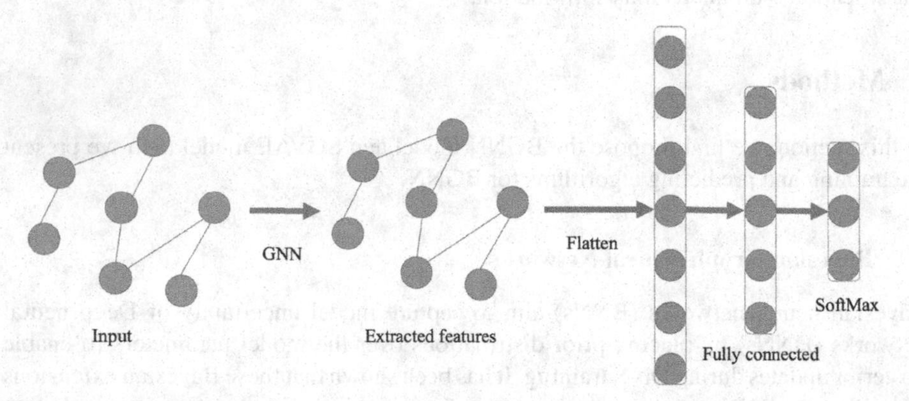

Fig. 1. The framework of the BGNN model for EEG emotion recognition. The inputs of the model are the EEG features extracted from multiple frequency bands, in which each EEG channel is represented as a node of the graph. The GNN layer can appear multiple time. After GNN we flatten the features into one-dimension vectors. The outputs are the predicted labels through softmax. The relu, dropout and batch normalization layers are not shown for simplicity.

2.2 Sparse Graph Variational Auto-encoder

In contrast to the $p(\mathcal{G}|\mathcal{G}_{obs})$ modeled by assortative mixed membership stochastic block model (a- MMSBM) in [16], we choose SGVAE which is adapted from [18] where they call it Deep Generative Latent Feature Relational Model (DGLFRM).

The SGVAE model is the graph extension of Variational Auto-encoder (VAE) [29] with the Gaussian prior distribution of latent features Z replaced by The Indian buffet process (IBP) [30]. There two advantages of SGVAE over a-MMSBM. The first one is the posterior distribution in SGVAE is modeled by Neural Networks which can be trained effectively while the inference in a-MMSBM is traditional variational inference (VI) which has poor scalability. The second is that the latent features in SGVAE have different sparsity according to our prior due to the property of IBP which means it can determine the number of latent communities automatically while in a-MMSBM we need other methods.

The plate notation of our model is demonstrated in Fig. 2.

Graph. We assume that the graph is given as an adjacency matrix $A \in \{0, 1\}^{N \times N}$, where N denotes the number of nodes. We assume $A_{nm} = 1$ if there exist a link between node

n and node m, and otherwise $A_{nm} = 0$. The graphs used in SGVAE are computed from Phase Locking Value (PLV) which is a statistic that can be used to investigate long range synchronization of neural activity from EEG data [14]. We compute PLV between every pairs of EEG channels. To get binary matrix, we use line search for the threshold that ensure about 30% elements in matrix are 1 [19].

Decoder. As in Fig. 2, we define the latent features of each EEG channels $Z_n = b_n \odot r_n \in \mathbb{R}^K$ where the $\odot$ operator is the element-wise product and K is the number of latent communities (Do not be confused with the extracted features Z in Sect. 2.1). The probability $p(A_{mn} = 1) = \sigma(Z_m \cdot Z_n) = \sigma(\sum_{i=1}^{K} Z_{mi} Z_{ni})$ where the σ is the sigmoid function $\sigma(x) = \frac{1}{1+e^{-x}}$. We found that it is more effective than the decoder used in [18]. The possible reason is as follows: the inner product reflects the similarity between two vectors, the bigger the product, the more probable the connection exists. The linear transform may destroy the similarity and introduce extra linear independence between Z.

As in [18], we model the binary vector $b_n \in \{0, 1\}^K$, denoting node-community memberships, using the stick-breaking construction of the IBP. The stick-breaking construction is given as follows

$$v_k \sim Beta(\alpha, 1), k = 1, \ldots, K \tag{3}$$

$$\pi_k = \prod_{j=1}^{k} v_j, b_{nk} \sim Bernoulli(\pi_k) \tag{4}$$

We further assume a Gaussian prior on membership strengths $p(r_n) = \mathcal{N}(0, I_{n\times n})$.

Encoder. In encoder we model the variational posteriors as parametric distributions whose parameters $c_{nk}, d_{nk}, \mu_{nk}, \sigma_{nk}$ are outputs of a GNN. We omit the encoder for π_n for it does not have extra parameters and is fully determined by c_n and d_n (through the stick-breaking construction of the IBP). In this paper we do not use the information from X and set it to identity matrix.

Inference. We define the factorized variational posterior $q_\varnothing(v, r)$ as

$$q_\varnothing(v_n) = Beta(v_n | c_n(A, X), d_n(A, X)) \tag{5}$$

$$q_\varnothing(r_n) = \mathcal{N}(\mu_n(A, X), diag(\sigma_n^2(A, X))) \tag{6}$$

The corresponding loss function $\mathcal{L}$ (negative of the evidence lower bound (ELBO)) is as follows

$$\mathcal{L} = \sum_{n=1}^{N} (KL[q_\varnothing(r_n) || p(r_n)] + KL[q_\varnothing(v_n) || p(v_n)]) - \sum_{n=1}^{N} \sum_{m=1}^{N} \mathbb{E}_q[log p_\theta(A_{mn} | Z_m, Z_n)] \tag{7}$$

More details of the models can be seen in [18].

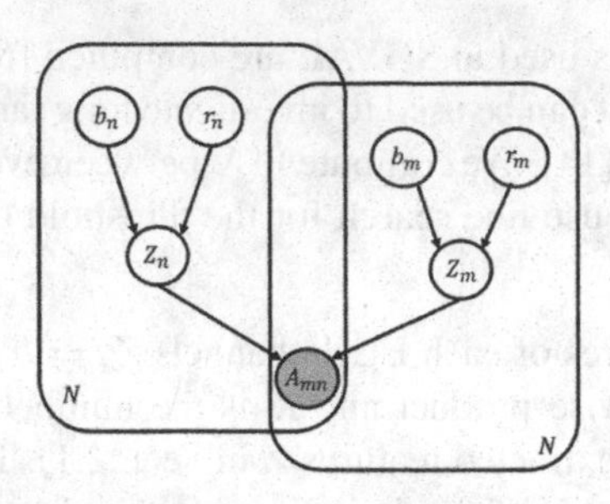

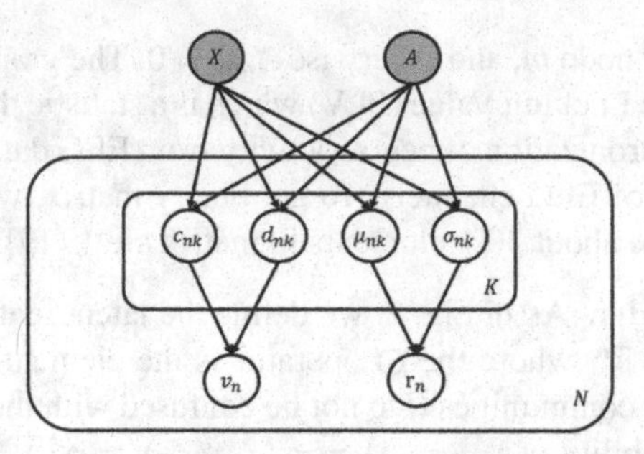

Fig. 2. (Left) The likelihood/decoder model in plate notation. (Right) The variational posterior/encoder model, defined by a GNN that takes as input the adjacency matrix A and node features X (if available) and outputs the parameters of the variational distributions of the latent variables. They gray nodes represent observed variables and white nodes represent hidden variables.

2.3 Algorithm for BGNN

The algorithm given by [16] has some defects. They trained the model on only one graph sampled from graph generative model which did not follows the Bayesian inference faithfully and need to re-train the model every time we want to make prediction.

In this section we propose the Algorithm 1 for training BGNN to overcome above defects and Algorithm 2 for predicting via trained BGNN. We use predictive entropy to represent the model uncertainty [20]

$$\widehat{\mathbb{H}}(y|x) = -\sum_{C} p\left(y^*|x^*, X, Y, \mathcal{G}_{obs}\right)\log\left(p\left(y^*|x^*, X, Y, \mathcal{G}_{obs}\right)\right) \tag{8}$$

where $p(y^*|x^*, X, Y, \mathcal{G}_{obs})$ can be approximate by (2).

Algorithm 1. BGNN-training

Input: Preprocessed EEG data X', labels Y
Output: The trained model
1. Compute adjacency matrix $\mathcal{G}_{obs}$ and extract features X from different band.
2. Train SGVAE on $\mathcal{G}_{obs}$.
3. **for** $i = 1 : epochs$ **do**
4. Sample $\mathcal{G}_i \sim p(\mathcal{G}|\mathcal{G}_{obs})$.
5. Train BGNN on $\mathcal{G}_i$ and X via concrete dropout [21].
6. **end for**

Algorithm 2. BGNN-predicting

Input: Extracted EEG data X^*
Output: The prediction labels Y^*
1. **for** $i = 1 : N_G$ **do**
2. Sample $\mathcal{G}_i \sim p(\mathcal{G}|\mathcal{G}_{obs})$.
3. **for** $j = 1 : S$ **do**
4. Sample $W_{i,j}$ and compute $p(Y^*_{i,j}|Z^*_{i,j}, W_{i,j})f(Z^*_{i,j}|X^*, \mathcal{G}_i, \theta)$
5. **end for**
6. **end for**
7. Approximate $p(Y^*|X^*, X, Y, \mathcal{G}_{obs})$ via (2)
8. Compute predictive entropy via (8)

3 Experiments

In this section, we present the datasets, classification settings and model settings in our experiments as well as classification results.

3.1 Datasets

We conduct experiments on a public datasets SEED [5]. The SEED dataset contains EEG data of 15 subjects (7 males and 8 females), which are collected via 62 EEG electrodes from the subjects when they are watching fifteen Chinese film clips with three types of emotions, i.e., negative, positive and neutral. To avoid making subjects fatigue, the whole experiment will not last for a long time and the duration of each film clip is about 4 min. The stimulus materials can be understood without explanation. Consequently, all the EEG signals will be categorized into one of three kinds of emotion states (positive, neutral and negative). The data collection lasted for 3 different periods corresponding to 3 sessions, and each session corresponds to 15 trials of EEG data such that there are totally 45 trials of EEG data for each subject. In addition, an additional subjective self-assessment for each subject is also carried out after the subjects watching the film clips in order to guarantee that the collected EEG data share the same emotion states as the film clips presented to the subjects.

3.2 Classification Settings

We closely follow prior studies [5] to conduct both subject-dependent and subject-independent classifications on SEED to evaluate our model. The differential entropy of all five bands (δ band, θ band, α band, β band, and γ band) are used for classification as in [5].

Our model is implemented via PyTorch [22] and PyTorch Geometric[23] two famous framework for deep learning and deep geometric learning.

For SGVAE, we use 5th order ChebConv with in_channels = 5 and out_channels = 16. We use identity matrix as input for we are not willing to reconstruct X. The prior hyper-parameter α is set to 2.

For BGNN, we use single layer ChebConv with in_channels = 5 and out_channels = 3 and MLP with size [3 * 62, 3 * 31, 3]. We use concrete dropout [21] to learn the model parameters and grid search the parameter space $10^{[-4:-2]}$ with a step of one for the hyper-parameters i.e. dropout regularizer and weight regularizer and [1 : 3] for the order of ChebConv. We use $N_G = 10, S = 10$ in subject-dependent classification and $N_G = 5, S = 5$ in subject-independent classification.

3.3 Results

Table 1 presents the classification results of our BGNN model and all baselines on SEED. The BGNN-CONF means we make prediction only on those model gives high confidence, e.g. the data that have predictive entropy less than the median of all data. Although our model dose not achieve the state-of-the-art results in normal prediction, we outperform all the others in data which we have high confidence in both classification setting which justifies the effectiveness of uncertainty information.

Table 1. Classification results (mean/std) on SEED. (left: subject-dependent) (right: subject-independent)

Model	All bands
SVM	83.99/09.92
DBN [5]	86.08/08.34
STRNN [24]	89.50/07.63
DGCNN [11]	90.40/08.49
BiDANN [25]	92.38/07.04
BiHDM [26]	93.12/06.06
RGNN [12]	**94.24/05.95**
BGNN(Ours)	89.15/08.47
BGNN-CONF	**96.37/05.17**

Model	All bands
SVM	56.73/16.29
T-SVM [27]	72.53/14.00
DGCNN [11]	79.95/09.02
DAN [28]	83.81/08.56
BiDANN-S [25]	84.14/06.87
BiHDM [26]	**85.40/07.53**
RGNN [12]	85.30/06.72
BGNN(Ours)	76.94/09.52
BGNN-CONF	**90.02/07.76**

4 Discussion

In this section, we conduct ablation study and analysis for our BGNN model. We also visualize the latent communities via our SGVAE.

4.1 Ablation Study

We conduct ablation study to justify the effectiveness of our proposed model. Table 2 reports the subject-dependent classification results on SEED. The GNN methods means non-Bayesian version of BGNN and MLP methods are similar to the one in BGNN but without topological information. To ensure fairness, we perform grid search in hyper-parameters like dropout rate, L2 regularizer and the order of ChebConv in GNN. The results show that topological information has some benefits but not significant in naïve implementation of GNN. The Bayesian version are least not harm to deep learning methods but can benefit from uncertainty information.

Table 2. Ablation study for subject-dependent classification results (mean/std) on SEED

Model	All data	CONF data
BGNN	**89.15/08.47**	**96.37/05.17**
GNN	88.13/09.19	–
MLP	87.72/09.04	–

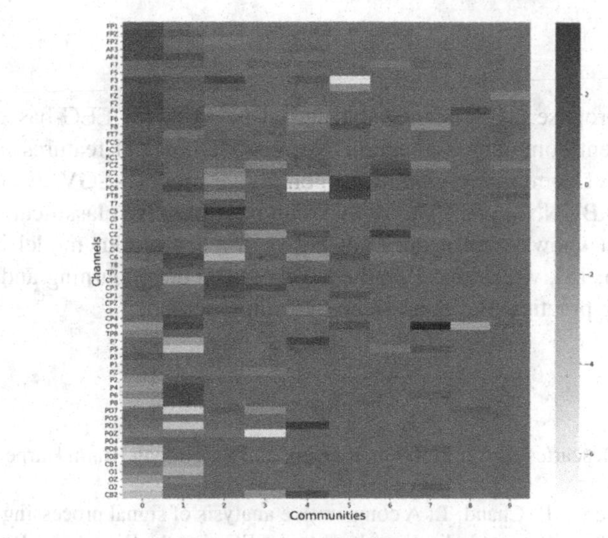

Fig. 3. Latent features from SGVAE

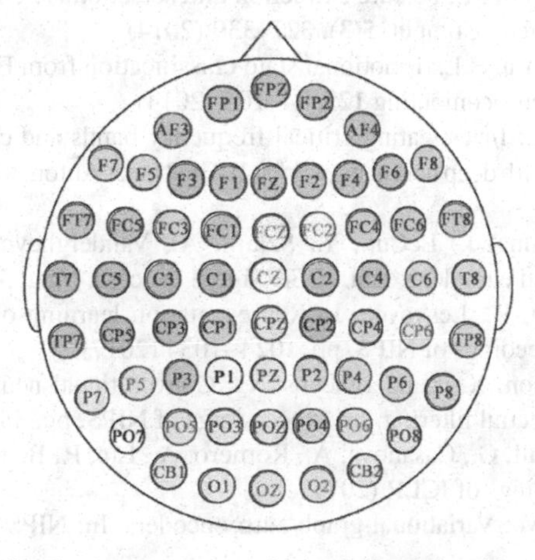

Fig. 4. The latent communities of channels with $\alpha = 2$

4.2 Latent Communities

In Fig. 3 we visualize the latent features encoded by SGVAE with $\alpha = 2, K = 10$. Each row represents a channel's membership, i.e. the strength it belongs to a community. We use the maximum component of each channel as its community membership and visualized in Fig. 4 (channels belong to the same community if they have the same color). It can be seen there exists some cluster in channels due to the physical distance or maybe functional connectivity.

5 Conclusions

In this paper, we propose a Bayesian graph neural network for EEG-based emotion recognition and latent community detection. We encode channel features into sparse latent space to detect community via a deep generative model—SGVAE and sample from it to train ours BGNN model. The BGNN can not only give classification results, more importantly, it knows what it does not know in a sense. Our model is inspired by the work [15, 16, 18], which make up the gap between deep learning and Bayesian modeling, providing practical Bayesian Deep Learning methods.

References

1. Damasio, A.R.: Descartes' Error: Emotion, Reason, and the Human Brain. Harper Perennial, New York (1995)
2. Khosla, A., Khandnor, P., Chand, T.: A comparative analysis of signal processing and classification methods for different applications based on EEG signals. Biocybern. Biomed. Eng. **40**, 649–690 (2020)
3. Jenke, R., Peer, A., Buss, M.: Feature extraction and selection for emotion recognition from EEG. IEEE Trans. Affect. Comput. **5**(3), 327–339 (2014)
4. Wang, X.W., Nie, D., Lu, B.L.: Emotional state classification from EEG data using machine learning approach. Neurocomputing **129**, 94–106 (2014)
5. Zheng, W.L., Lu, B.L.: Investigating critical frequency bands and channels for EEG-based emotion recognition with deep neural networks. IEEE Trans. Auton. Ment. Dev. **7**(3), 162–175 (2015)
6. Bronstein, M.M., Bruna, J., LeCun, Y., Szlam, A., Vandergheynst, P.: Geometric deep learning: going beyond euclidean data. IEEE Signal Process. Mag. **34**(4), 18–42 (2017)
7. Hamilton, W.L., Ying, R., Leskovec, J.: Representation learning on graphs: Methods and applications. In: Proceedings of NIPS, pp. 1024–1034 (2017)
8. Defferrard, M., Bresson, X., Vandergheynst, P.: Convolutional neural networks on graphs with fast localized spectral filtering. In: Proceedings of NIPS, pp. 3844–3852 (2016)
9. Velickovic, P., Cucurull, G., Casanova, A., Romero, A., Lio, P., Bengio, Y.: Graph attention networks. In: Proceedings of ICLR (2017)
10. Kipf, T.N., Welling, M.: Variational graph auto-encoders. In: NIPS Workshop on Bayesian Deep Learning (2016)
11. Song, T., Zheng, W., Song, P., Cui, Z.: EEG emotion recognition using dynamical graph convolutional neural networks. IEEE Trans. Affect. Comput. **11**(3), 532–541 (2020)
12. Zhong, P.X., Wang, D., Miao, C.Y.: EEG-based emotion recognition using regularized graph neural networks. IEEE Transactions on Affective Computing (in press)

13. Ding, Y., Robinson, N., Zeng, Q.H., Guan, C.T.: LGGNet: learning from Local-global-graph representations for brain-computer interface. arXiv preprint arXiv:2105.02786 (2021)
14. Wang, Z.M., Tong, Y., Heng, X.: Phase-locking value based graph convolutional neural networks for emotion recognition. IEEE Access **7**, 93711–93722 (2019)
15. Gal, Y., Ghahramani, Z.: Dropout as a Bayesian approximation: representing model uncertainty in deep learning. In: Proceedings of International Conference Machine Learning (2016)
16. Zhang, Y.X., Pal, S., Coates, M., Ustebay, D.: Bayesian graph convolutional neural networks for semi-supervised classification. In: Proceedings of Conference of AAAI (2019)
17. Hasanzadeh, A., Hajiramezanali, E., et al.: Bayesian graph neural networks with adaptive connection sampling. In: Proceedings of International Conference Machine Learning, PMLR, vol. 119 (2020)
18. Mehta, N., Duke, L.C., Rai, P.: Stochastic blockmodels meet graph neural networks. In: Proceedings International Conference Machine Learning, PMLR, vol. 97, pp. 4466–4474 (2019)
19. Achard, S., Bullmore, E.: Efficiency and cost of economical brain functional networks. PLoS Comput. Biol. **3**(2), e17 (2007)
20. Mukhoti, J., Gal, Y.: Evaluating Bayesian deep learning methods for semantic segmentation. arXiv preprint arXiv:1811.12709 (2019)
21. Gal, Y., Hron, J., Kendall, A.: Concrete dropout. arXiv preprint arXiv:1705.07832 (2017)
22. Paszke, A., Gross, S., et al.: PyTorch: an imperative style, high-performance deep learning library. In Conference NeurIPS (2019)
23. Fey, M., Lenssen, J.E.: Fast graph representation learning with PyTorch geometric. In: ICLR (2019)
24. Zhang, T., Zheng, W., Cui, Z., Zong, Y., Li, Y.: Spatial-temporal recurrent neural network for emotion recognition. IEEE Trans. Cybern. **99**, 1–9 (2018)
25. Li, Y., Zheng, W., Zong, Y., Cui, Z., Zhang, T., Zhou, X.: A bi-hemisphere domain adversarial neural network model for EEG emotion recognition. IEEE Transactions on Affective Computing (2018, in press)
26. Li, Y., et al.: A novel bi-hemispheric discrepancy model for EEG emotion recognition. arXiv preprint arXiv:1906.01704 (2019)
27. Collobert, R., Sinz, F., Weston, J., Bottou, L.: Large scale transductive SVM. J. Mach. Learn. Res. **7**, 1687–1712 (2016)
28. Li, H., Jin, Y.-M., Zheng, W.-L., Bao-Liang, L.: Cross-subject emotion recognition using deep adaptation networks. In: Cheng, L., Leung, A.C.S., Ozawa, S. (eds.) ICONIP 2018. LNCS, vol. 11305, pp. 403–413. Springer, Cham (2018). https://doi.org/10.1007/978-3-030-04221-9_36
29. Kingma, D.P., Welling, M.: Auto-encoding variational Bayes. arXiv preprint arXiv:1312.6114 (2013)
30. Griffiths, T.L., Ghahramani, Z.: The Indian buffet process: an introduction and review. J. Mach. Learn. Res. **12**, 1185–1224 (2011)

ViTBIS: Vision Transformer for Biomedical Image Segmentation

Abhinav Sagar(✉)

Vellore Institute of Technology, Vellore, India

Abstract. In this paper, we propose a novel network named Vision Transformer for Biomedical Image Segmentation (ViTBIS). Our network splits the input feature maps into three parts with 1×1, 3×3 and 5×5 convolutions in both encoder and decoder. Concat operator is used to merge the features before being fed to three consecutive transformer blocks with attention mechanism embedded inside it. Skip connections are used to connect encoder and decoder transformer blocks. Similarly, transformer blocks and multi scale architecture is used in decoder before being linearly projected to produce the output segmentation map. We test the performance of our network using Synapse multi-organ segmentation dataset, Automated cardiac diagnosis challenge dataset, Brain tumour MRI segmentation dataset and Spleen CT segmentation dataset. Without bells and whistles, our network outperforms most of the previous state of the art CNN and transformer based models using Dice score and the Hausdorff distance as the evaluation metrics.

1 Introduction

Deep Convolutional Neural Networks has been highly successful in medical image segmentation. U-Net (Ronneberger et al. 2015) based architectures use a symmetric encoder-decoder network with skip-connections. The limitation of CNN-based approach is that it is unable to model long-range relation, due to the regional locality of convolution operations. To tackle this problem, self attention mechanism was proposed (Schlemper et al. 2019) and (Wang et al. 2018). Still, the problem of capturing multi-scale contextual information was not solved which leads not so accurate segmentation of structures with variable shapes and scales (e.g. brain lesions with different sizes). An alternative technique using Transformers are better suited at modeling global contextual information.

Vision Transformer (ViT) (Dosovitskiy et al. 2020) splits the image into patches and models the correlation between these patches as sequences with Transformer, achieving better speed-performance trade-off on image classification than previous state of the art image recognition methods. DeiT (Touvron et al. 2020) proposed a knowledge distillation method for training Vision Transformers. An extensive study was done by (Bakas et al. 2018) to find the best algorithm for segmenting tumours in brain. Medical images from CT and MRI are in 3 dimensions, thus making volumetric segmentation important. Çiçek et

C. Oyarzun Laura et al. (Eds.): CLIP/DCL/LL-COVID/PPML 2021, LNCS 12969, pp. 34–45, 2021.
https://doi.org/10.1007/978-3-030-90874-4_4

al. (2016) tackled this problem using 3d U-Net. Densely-connected volumetric convnets was used (Yu et al. 2017) to segment cardiovascular images. A comprehensive study to evaluate segmentation performance using Dice score and Jaccard index was done by (Eelbode et al. 2020).

2 Related Work

2.1 Convolutional Neural Network

Earlier work for medical image segmentation used some variants of the original U-shaped architecture (Ronneberger et al. 2015). Some of these were Res-UNet (Xiao et al. 2018), Dense-UNet (Li et al. 2018) and U-Net++ (Zhou et al. 2018). These architectures are quite successful for various kind of problems in the domain of medical image segmentation.

2.2 Attention Mechanism

Self Attention mechanism (Wang et al. 2018) has been used successfully to improve the performance of the network. (Schlemper et al. 2019) used skip connections with additive attention gate in U-shaped architecture to perform medical image segmentation. Attention mechanism was first used in U-Net (Oktay et al. 2018) for medical image segmentation. A multi-scale attention network (Fan et al. 2020) was proposed in the context of biomedical image segmentation. (Jin et al. 2020) used a hybrid deep attention-aware network to extract liver and tumor in ct scans. Attention module was added to U-Net module to exploit full resolution features for medical image segmentation (Li et al. 2020). A similar work using attention based CNN was done by (Liu et al. 2020) in the context of schemic stroke disease. A multi scale self guided attention network was used to achieve state of the art results (Sinha and Dolz 2020) for medical image segmentation.

2.3 Transformers

Transformer first proposed by (Vaswani et al. 2017) have achieved state of the art performance on various tasks. Inspired by it, Vision Transformer (Dosovitskiy et al. 2020) was proposed which achieved better speed-accuracy tradeoff for image recognition. To improve this, Swin Transformer (Liu et al. 2021) was proposed which outperformed previous networks on various vision tasks including image classification, object detection and semantic segmentation. (Chen et al. 2021), (Valanarasu et al. 2021) and (Hatamizadeh et al. 2021) individually proposed methods to integrate CNN and transformers into a single network for medical image segmentation. Transformer along with CNN are applied in multi-modal brain tumor segmentation (Wang et al. 2021) and 3D medical image segmentation (Xie et al. 2021).

Our main contributions can be summarized as:

- We propose a novel network incorporating attention mechanism in transformer architecture along with multi scale module name ViTBIS in the context of medical image segmentation.
- Our network outperforms previous state of the art CNN based as well as transformer based architectures on various datasets.
- We present the ablation study showing our network performance is generalizable hence can be incorporated to tackle other similar problems.

2.4 Background

Suppose an image is given $x \in R^{H \times W \times C}$ with a spatial resolution of $H \times W$ and C number of channels. The goal is to predict the pixel-wise label of size $H \times W$ for each image. We start by performing tokenization by reshaping the input x into a sequence of flattened 2D patches $x_p^i \in R(i = 1, .., N)$, where each patch is of size $P \times P$ and $N = (H \times W)/P^2$ is the number of patches present in the image. We convert the vectorized patches x_p into a latent D-dimensional embedding space using a linear projection vector. We use patch embeddings to make sure the positional information is present as shown below:

$$\mathbf{z}_0 = [\mathbf{x}_p^1\mathbf{E}; \mathbf{x}_p^2\mathbf{E}; \cdots; \mathbf{x}_p^N\mathbf{E}] + \mathbf{E}_{pos} \tag{1}$$

where $E \in R^{(P^2C)} \times D$ denotes the patch embedding projection, and $E_{pos} \in R^{N \times D}$ denotes the position embedding.

After the embedding layer, we use multi scale context block followed by a stack of transformer blocks (Dosovitskiy et al. 2020) made up of multiheaded self-attention (MSA) and multilayer perceptron (MLP) layers as shown in Eq. 2 and Eq. 3 respectively:

$$\mathbf{z}_i' = \mathrm{MSA}\left(\mathrm{Norm}\left(\mathbf{z}_{i-1}\right)\right) + \mathbf{z}_{i-1} \tag{2}$$

$$\mathbf{z}_i = \mathrm{MLP}\left(\mathrm{Norm}\left(\mathbf{z}_i'\right)\right) + \mathbf{z}_i' \tag{3}$$

where Norm represents layer normalization, MLP is made up of two linear layers and i is the individual block. A MSA block is made up of n self-attention (SA) heads in parallel. The structure of Transformer layer used in this work is illustrated in Fig. 1:

3 Method

3.1 Dataset

1. Synapse multi-organ segmentation dataset - We use 30 abdominal CT scans in the MICCAI 2015 Multi-Atlas Abdomen Labeling Challenge, with 3779 axial contrast-enhanced abdominal clinical CT images in total.

2. Brain Tumor Segmentation dataset - 3D MRI dataset used in the experiments is provided by the BraTS 2019 challenge (Menze et al. 2014) and (Bakas et al. 2018).

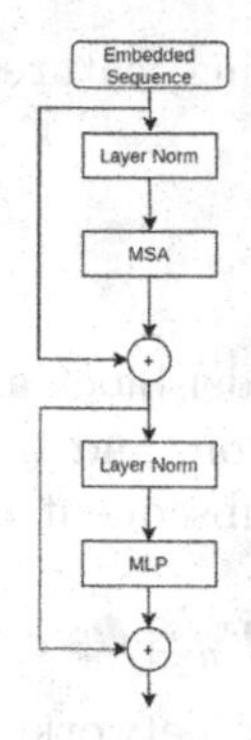

Fig. 1. Schematic of the Transformer layer used in this work.

3.2 Network Architecture

The output sequence of Transformer $z_L \in R^{d \times N}$ is first reshaped to $d \times H/8 \times W/8 \times D/8$. A convolution block is used to reduce the channel dimension from d to K. This helps in reducing the computational complexity. Upsampling operations and successive convolution blocks are the used to get back a full resolution segmentation result $R \in R^{H \times W \times D}$. Skip-connections are used to fuse the encoder features with the decoder by concatenation to get more contextual information. In the encoder part, the input image is split into patches and fed into linear embedding layer. The feature map is splitted into N parts along with the channel dimension. The individual features are fused before being passed to the transformer blocks. The decoder block is comprised of transformer blocks followed by a similar split and concat operator. Linear projection is used on the feature maps to produce the segmentation map. Skip connections are used between the encoder and decoder transformer blocks to provide an alternative path for the gradient to flow thus speeding up the training process.

Two different types of convolutional operations are applied to the encoder features F_{en} to generate the feature maps $F_1 \in R_1$ and $F_2 \in R^{c \times h \times w}$ respectively. Subsequently, F is reshaped into the matrixes of feature maps F_1 and F_2. Then, a matrix multiplication operation with softmax normalization is performed in the permuted version of M and N, resulting in the position attention map $B \in R(h \times w) \times (h \times w)$, which can be defined as:

$$\boldsymbol{B}_{i,j} = \frac{\exp\left(\boldsymbol{M}_i \cdot \boldsymbol{N}_j\right)}{\sum_{i=1}^{n} \exp\left(\boldsymbol{M}_i \cdot \boldsymbol{N}_j\right)} \tag{4}$$

where $B_{i,j}$ measures the impact of ith position on j^{th} position and n = h × w is the number of pixels. After that, W is multiplied by the permuted version of B, and the resulting feature at each position can be formulated as:

$$GSA(\boldsymbol{M}, \boldsymbol{N}, \boldsymbol{W})_j = \sum_{i=1}^{n} (\boldsymbol{B}_{i,j} \boldsymbol{W}_j) \tag{5}$$

Similarly, we reshape the resulting features to generate the final output of our vision transformer.

3.3 Residual Connection

The input feature maps of each decoder block are up-sampled to the resolution of outputs through bilinear interpolation, and then concatenated with the output feature maps as the inputs of the subsequent block, which is defined as:

$$\boldsymbol{F}_n = f_n\left(\boldsymbol{F}_{n-1}\right) \oplus v_n\left(\boldsymbol{F}_{n-1}\right) \tag{6}$$

The detailed architecture of our network as well as the intermediate skip-connections is shown in Fig. 2:

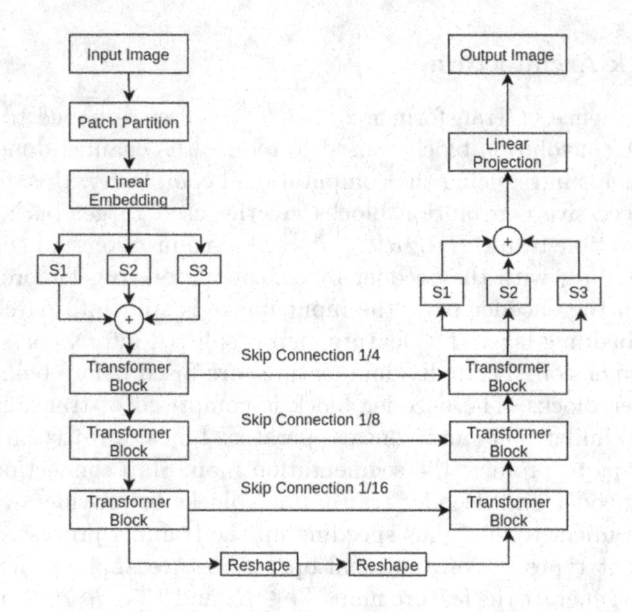

Fig. 2. Overview of our model architecture. Output sizes demonstrated for patch dimension N = 16 and embedding size C = 768. We extract sequence representations of different layers in the transformer and merge them with the decoder using skip connections.

Similar to the previous works (Hu et al. 2019), self-attention is computed as defined below:

$$\text{MSA}\ (Q, K, V) = \text{Sof}\,tMax\left(\frac{QK^T}{\sqrt{d}} + B\right)V \tag{7}$$

where $Q, K, V \in R^{M^2 \times d}$ denote the query, key and value matrices. M^2 and d denotes the number of patches in a window and the dimension of the query. The values in B are taken from the random bias matrix denoted by $B \in R^{(2M-1)\times(2M+1)}$

The output of MSA is defined below:

$$\text{TMSA}(\mathbf{z}) = [\text{MSA}_1(z); \text{MSA}_2(z); \ldots; \text{MSA}_n(z)]\, \mathbf{W}_{tmsa} \tag{8}$$

Where W_{tmsa} represents the learnable weight matrices of different heads (SA).

3.4 Loss Function

Commonly used Binary Cross Entropy and Dice Loss terms are used for training our network as defined in Eq. 9 and Eq. 10 respectively:

$$\mathcal{L}_{BCE} = \sum_{i=1}^{t} (y_i \log(p_i) + (1 - y_i) \log(1 - p_i)) \tag{9}$$

$$\mathcal{L}_{\text{Dice}} = 1 - \frac{\sum_{i=1}^{t} y_i p_i + \varepsilon}{\sum_{i=1}^{t} y_i + p_i + \varepsilon} \tag{10}$$

where t is the total number of pixels in each image, y_i represents the ground-truth value of the ith pixel, p_i the confidence score of the ith pixel in prediction results. The above two loss functions can be combined to give:

$$L_{total} = L_{BCE} + L_{Dice} \tag{11}$$

The complete loss function is a combination of dice and cross entropy terms which is calculated in voxel-wise manner as defined below:

$$L_{total} = 1 - \alpha\frac{2}{J}\sum_{j=1}^{J}\frac{\sum_{i=1}^{I} G_{i,j}Y_{i,j}}{\sum_{i=1}^{I} G_{i,j}^2 + \sum_{i=1}^{I} Y_{i,j}^2} + \beta\frac{1}{I}\sum_{i=1}^{I}\sum_{j=1}^{J} G_{i,j}\log Y_{i,j} \tag{12}$$

where I is the number of voxels, J is the number of classes, $Y_{i,j}$ and $G_{i,j}$ denote the probability output and one-hot encoded ground truth for voxel i of class j. In our experiment, $\alpha = \beta = 0.5$, and $\epsilon = 0.0001$.

3.5 Evaluation Metrics

The segmentation accuracy is measured by the Dice score and the Hausdorff distance (95%) metrics for enhancing tumor region (ET), regions of the tumor core (TC), and the whole tumor region (WT).

3.6 Implementation Details

Our model is trained using Pytorch deep learning framework. The learning rate and weight decay values used are 0.00015 and 0.005, respectively. We use batch size value of 16 and ADAM optimizer to train our model. We use a random crop of $128 \times 192 \times 192$ and mean normalization to prepare our model input. The input image size and patch size are set as 224×224 and 4, respectively. As a model input, we use the 3D voxel by cropping the brain region. The following data augmentation techniques are applied:

1. Random cropping of the data from $240 \times 240 \times 155$ to $128 \times 128 \times 128$ voxels;
2. Flipping across the axial, coronal and sagittal planes by a probability of 0.5
3. Random Intensity shift between [−0.05, 0.05] and scale between [0.5, 1.0].

4 Results

We report the average DSC and average Hausdorff Distance (HD) on 8 abdominal organs (aorta, gallbladder, spleen, left kidney, right kidney, liver, pancreas, spleen, stomach) with a random split of 20 samples in training set and 10 sample for validation set using Synapse multi-organ CT dataset in Table 1. Our network clearly outperforms previous state of the art CNN as well as transformer networks.

Table 1. Comparison on the Synapse multi-organ CT dataset (average dice score %, average hausdorff distance in mm, and dice score % for each organ). The best results are highlighted in bold.

Encoder	Decoder	DSC	HD	Aorta	GB	Kid(L)	Kid(R)	Liver	Panc	Spleen	Stomach
V-Net	V-Net	68.81	–	75.34	51.87	77.10	80.75	87.84	40.05	80.56	56.98
DARR	DARR	69.77	–	74.74	53.77	72.31	73.24	94.08	54.18	89.90	45.96
R50	U-Net	74.68	36.87	84.18	62.84	79.19	71.29	93.35	48.23	84.41	73.92
R50	AttnUNet	75.57	36.97	55.92	63.91	79.20	72.71	93.56	49.37	87.19	74.95
ViTBIS	None	61.50	39.61	44.38	39.59	67.46	62.94	89.21	43.14	75.45	69.78
ViTBIS	CUP	67.86	36.11	70.19	45.10	74.70	67.40	91.32	42.00	81.75	70.44
R50-ViTBIS	CUP	71.29	32.87	73.73	55.13	75.80	72.20	91.51	45.99	81.99	73.95
TransUNet	TransUNet	77.48	31.69	87.23	63.13	81.87	77.02	94.08	55.86	85.08	75.62
SwinUnet	SwinUnet	79.13	21.55	85.47	66.53	83.28	79.61	94.29	56.58	90.66	76.60
ViTBIS	ViTBIS	**80.45**	**21.24**	**86.41**	**66.80**	**83.59**	**80.12**	**94.56**	**56.90**	**91.28**	**76.82**

We conduct the five-fold cross-validation evaluation on the BraTS 2019 training set. The quantitative results is presented in Table 2. Our network again outperforms previous state of the art CNN as well as transformer networks using most of the evaluation metrics except Hausdorff distance on ET and WT.

We compare the performance of our model against CNN based networks for the task of brain tumour segmentation in Table 3. Again, our network outperforms previous state of the art CNN as well as transformer networks.

Table 2. Comparison on the BraTS 2019 validation set. DS represents Dice score and HD repesents Hausdorff distance. The best results are highlighted in bold.

Method	ET(DS%)	WT(DS%)	TC(DS%)	ET(HD mm)	WT(HD mm)	TC(HD mm)
3D U-Net	70.86	87.38	72.48	5.062	9.432	8.719
V-Net	73.89	88.73	76.56	6.131	6.256	8.705
KiU-Net	73.21	87.60	73.92	6.323	8.942	9.893
Attention U-Net	75.96	88.81	77.20	5.202	7.756	8.258
Li et al.	77.10	88.60	81.30	6.033	6.232	7.409
TransBTS w/o TTA	78.36	88.89	81.41	5.908	7.599	7.584
TransBTS w/ TTA	78.93	90.00	81.94	**3.736**	**5.644**	6.049
ViTBIS	**79.24**	**90.28**	**82.23**	3.706	5.621	**7.129**

Table 3. Cross validation results of brain tumour Segmentation task. DSC1, DSC2 and DSC3 denote average dice scores for the Whole Tumour (WT), Enhancing Tumour (ET) and Tumour Core (TC) across all folds. For each split, average dice score of three classes are used. The best results are highlighted in bold.

Fold	Split-1	Split-2	Split-3	Split-4	Split-5	DSC1	DSC2	DSC3	Avg.
VNet	64.83	67.28	65.23	65.2	66.34	75.96	54.99	66.38	65.77
AHNet	65.78	69.31	65.16	65.05	67.84	75.8	57.58	66.50	66.63
Att-UNet	66.39	70.18	65.39	66.11	67.29	75.29	57.11	68.81	67.07
UNet	67.20	69.11	66.84	66.95	68.16	75.03	57.87	70.06	67.65
SegResNet	69.62	71.84	67.86	68.52	70.43	76.37	59.56	73.03	69.65
ViTBIS	**70.92**	**73.84**	**71.05**	**72.29**	**72.43**	**79.52**	**60.90**	**76.11**	**71.98**

In Table 4, We compare the performance of our network against previous state of the art for the task of spleen segmentation. Except on Split-4 and Split-5, our network outperforms both state of the art CNN and transformer networks.

The visualization of the validation set prediction is illustrated in Fig. 3:

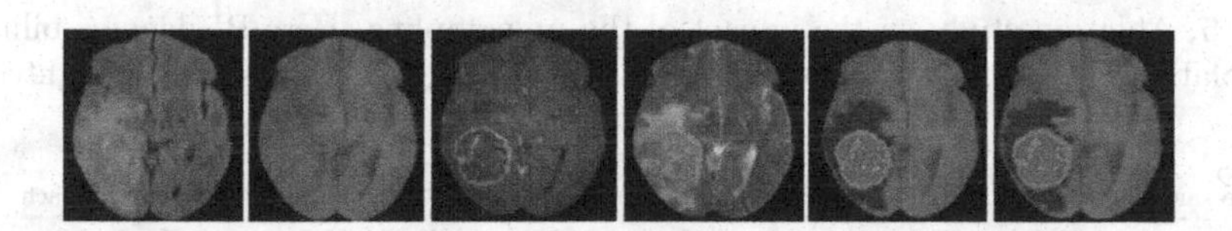

Fig. 3. All the four modalities of the brain tumor visualized with the Ground-Truth and Predicted segmentation of tumor sub-regions for BraTS 2019 crossvalidation dataset. Red label: Necrosis, yellow label: Edema and Green label: Edema. (Color figure online)

The segmentation results of our model on the Synapse multi-organ CT dataset is shown in Fig. 4:

Table 4. Cross validation results of spleen segmentation task. For each split, we provide the average dice score of fore-ground class. The best results are highlighted in bold.

Fold	Split-1	Split-2	Split-3	Split-4	Split-5	Avg.
VNet	94.78	92.08	95.54	94.73	95.03	94.43
AHNet	94.23	92.10	94.56	94.39	94.11	93.87
Att-UNet	93.16	92.59	95.08	94.75	95.81	94.27
UNet	92.83	92.83	95.76	95.01	96.27	94.54
SegResNet	95.66	92.00	95.79	94.19	95.53	94.63
UNETR	95.95	94.01	96.37	**95.89**	**96.91**	95.82
ViTBIS	**96.14**	**94.52**	**96.52**	95.76	96.78	**96.14**

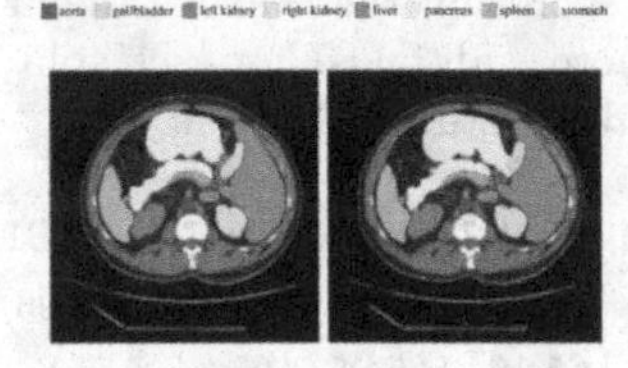

Fig. 4. The segmentation results of our network on the Synapse multi-organ CT dataset. Left depicts ground truth, while the right one depicts predicted segmentation from our network.

4.1 Ablation Studies

We conduct the experiments of our model with bilinear interpolation and transposed convolution on Synapse multi-organ CT dataset as shown in Table 5. The experiment shows that our network using transposed convolution layer achieves better segmentation accuracy.

Table 5. Ablation study on the impact of the up-sampling. Here BI denotes bilinear interpolation, TC denotes transposed convolution. The best results are highlighted in bold.

Up-sampling	DSC	Aorta	Gallbladder	Kidney(L)	Kidney(R)	Liver	Pancreas	Spleen	Stomach
BI	77.24	82.04	67.18	80.52	73.79	94.05	55.74	86.71	72.50
TC	**78.53**	**84.55**	**68.02**	**82.46**	**74.41**	**94.59**	**55.91**	**89.25**	**73.96**

We explore our network at various model scales (i.e. depth (L) and embedding dimension (d)) using BraTS 2019 validation dataset. We show ablation study to verify the impact of Transformer scale on the segmentation performance. Our network with d = 384 and L = 4 achieves the best scores of ET, WT and TC. Increasing the depth and decreasing the embedding dimension gives better results. However, the impact of depth on performance is much more than that of embedding dimension as shown in Table 6:

Table 6. Ablation study demonstrating the effect of depth and embedding dimension on our vision transformer using BraTS 2019 validation dataset. DS represents Dice score. The best results are highlighted in bold.

Depth (L)	Embedding dim (d)	ET(DS%)	WT(DS%)	TC(DS%)
1	384	69.24	84.16	70.18
1	512	69.05	83.87	69.92
2	384	70.59	84.88	72.51
2	512	70.13	84.15	71.99
4	384	**72.06**	**85.39**	**73.67**
4	512	71.55	85.06	73.05

Using the set of ablation studies, it can be inferred that the performance of our network is generalizable.

5 Conclusions

Biomedical image segmentation is a challenging problem in medical imaging. Recently deep learning methods leveraging both CNN and transformer based architectures have been highly successful in this domain. In this paper, we propose a novel network named Vision Transformer (ViTBIS) for Biomedical Image Segmentation. We use multi scale mechanism to split the features employing different convolutions and concatenating those individual feature maps produced before being passed to transformer blocks in encoder. The decoder also uses similar mechanism with skip connections connecting the encoder and decoder transformer blocks. The output feature map after split and concat operator is passed through a linear projection block to produce the output segmentation map. Using Dice Score and the Hausdorff Distance on multiple datasets, our network outperforms most of the previous CNN as well as transformer based architectures. In the future, we would like to use multi scale vision transformer to tackle other problems in computer vision like depth estimation.

References

Bakas, S., et al.: Identifying the best machine learning algorithms for brain tumor segmentation, progression assessment, and overall survival prediction in the brats challenge. arXiv preprint arXiv:1811.02629 (2018)

Cao, H., et al.: Swin-Unet: Unet-like pure transformer for medical image segmentation. arXiv preprint arXiv:2105.05537 (2021)

Chen, J., et al.: TransUNet: transformers make strong encoders for medical image segmentation. arXiv preprint arXiv:2102.04306 (2021)

Çiçek, Ö., Abdulkadir, A., Lienkamp, S.S., Brox, T., Ronneberger, O.: 3D U-Net: learning dense volumetric segmentation from sparse annotation. In: Ourselin, S., Joskowicz, L., Sabuncu, M.R., Unal, G., Wells, W. (eds.) MICCAI 2016. LNCS, vol. 9901, pp. 424–432. Springer, Cham (2016). https://doi.org/10.1007/978-3-319-46723-8_49

Dosovitskiy, A., et al.: An image is worth 16x16 words: transformers for image recognition at scale. arXiv preprint arXiv:2010.11929 (2020)
Eelbode, T., et al.: Optimization for medical image segmentation: theory and practice when evaluating with dice score or Jaccard index. IEEE Trans. Med. Imaging **39**(11), 3679–3690 (2020)
Fan, T., Wang, G., Li, Y., Wang, H.: MA-Net: a multi-scale attention network for liver and tumor segmentation. IEEE Access **8**, 179656–179665 (2020)
Fu, S., et al.: Domain adaptive relational reasoning for 3D multi-organ segmentation. In: Martel, A.L. (ed.) MICCAI 2020. LNCS, vol. 12261, pp. 656–666. Springer, Cham (2020). https://doi.org/10.1007/978-3-030-59710-8_64
Gibson, E., et al.: Automatic multi-organ segmentation on abdominal CT with dense v-networks. IEEE Trans. Med. Imaging **37**(8), 1822–1834 (2018)
Hatamizadeh, A., Yang, D., Roth, H., Xu, D.: UNETR: transformers for 3D medical image segmentation. arXiv preprint arXiv:2103.10504 (2021)
Hu, H., Zhang, Z., Xie, Z., Lin, S.: Local relation networks for image recognition. In: Proceedings of the IEEE/CVF International Conference on Computer Vision, pp. 3464–3473 (2019)
Isensee, F., Jaeger, P.F., Kohl, S.A., Petersen, J., Maier-Hein, K.H.: nnU-NET: a self-configuring method for deep learning-based biomedical image segmentation. Nat. Methods **18**(2), 203–211 (2021)
Jin, Q., Meng, Z., Sun, C., Cui, H., Su, R.: RA-Unet: a hybrid deep attention-aware network to extract liver and tumor in CT scans. Front. Bioeng. Biotechnol. **8**, 1471 (2020)
Li, C., et al.: ANU-Net: attention-based nested U-Net to exploit full resolution features for medical image segmentation. Comput. Graph. **90**, 11–20 (2020)
Li, X., Chen, H., Qi, X., Dou, Q., Fu, C.-W., Heng, P.-A.: H-DenseUNet: hybrid densely connected UNet for liver and tumor segmentation from CT volumes. IEEE Trans. Med. Imaging **37**(12), 2663–2674 (2018)
Liu, L., Kurgan, L., Wu, F.-X., Wang, J.: Attention convolutional neural network for accurate segmentation and quantification of lesions in ischemic stroke disease. Med. Image Anal. **65**, 101791 (2020)
Liu, Z., et al.: Swin transformer: hierarchical vision transformer using shifted windows. arXiv preprint arXiv:2103.14030 (2021)
Menze, B.H., et al.: The multimodal brain tumor image segmentation benchmark (brats). IEEE Trans. Med. Imaging **34**(10), 1993–2024 (2014)
Myronenko, A.: 3D MRI brain tumor segmentation using autoencoder regularization. In: Crimi, A., Bakas, S., Kuijf, H., Keyvan, F., Reyes, M., van Walsum, T. (eds.) BrainLes 2018. LNCS, vol. 11384, pp. 311–320. Springer, Cham (2019). https://doi.org/10.1007/978-3-030-11726-9_28
Ni, J., Wu, J., Tong, J., Chen, Z., Zhao, J.: GC-Net: global context network for medical image segmentation. Comput. Methods Programs Biomed. **190**, 105121 (2020)
Oktay, O., et al.: Attention U-Net: learning where to look for the pancreas. arXiv preprint arXiv:1804.03999 (2018)
Parmar, N., et al.: Image transformer. In: International Conference on Machine Learning, pp. 4055–4064. PMLR (2018)
Ronneberger, O., Fischer, P., Brox, T.: U-Net: convolutional networks for biomedical image segmentation. In: Navab, N., Hornegger, J., Wells, W.M., Frangi, A.F. (eds.) MICCAI 2015. LNCS, vol. 9351, pp. 234–241. Springer, Cham (2015). https://doi.org/10.1007/978-3-319-24574-4_28
Sagar, A.: Bayesian multi scale neural network for crowd counting. arXiv preprint arXiv:2007.14245 (2020a)

Sagar, A.: Monocular depth estimation using multi scale neural network and feature fusion. arXiv preprint arXiv:2009.09934 (2020b)

Sagar, A.: DMSANet: dual multi scale attention network. arXiv preprint arXiv:2106.08382 (2021)

Sagar, A., Soundrapandiyan, R.: Semantic segmentation with multi scale spatial attention for self driving cars. arXiv preprint arXiv:2007.12685 (2020)

Schlemper, J., et al.: Attention gated networks: learning to leverage salient regions in medical images. Med. Image Anal. **53**, 197–207 (2019)

Simpson, A.L., et al.: A large annotated medical image dataset for the development and evaluation of segmentation algorithms. arXiv preprint arXiv:1902.09063 (2019)

Sinha, A., Dolz, J.: Multi-scale self-guided attention for medical image segmentation. IEEE J. Biomed. Health Inform. **25**, 121–130 (2020)

Touvron, H., Cord, M., Douze, M., Massa, F., Sablayrolles, A., Jégou, H.: Training data-efficient image transformers & distillation through attention. arXiv preprint arXiv:2012.12877 (2020)

Valanarasu, J.M.J., Oza, P., Hacihaliloglu, I., Patel, V.M.: Medical transformer: gated axial-attention for medical image segmentation. arXiv preprint arXiv:2102.10662 (2021)

Vaswani, A., et al.: Attention is all you need. arXiv preprint arXiv:1706.03762 (2017)

Wang, W., Chen, C., Ding, M., Li, J., Yu, H., Zha, S.: TransBTS: multimodal brain tumor segmentation using transformer. arXiv preprint arXiv:2103.04430 (2021)

Wang, X., Girshick, R., Gupta, A., He, K.: Non-local neural networks. In: Proceedings of the IEEE Conference on Computer Vision and Pattern Recognition, pp. 7794–7803 (2018)

Xiao, X., Lian, S., Luo, Z., Li, S.: Weighted res-UNet for high-quality retina vessel segmentation. In: 2018 9th International Conference on Information Technology in Medicine and Education (ITME), pp. 327–331. IEEE (2018)

Xie, Y., Zhang, J., Shen, C., Xia, Y.: COTR: Efficiently bridging CNN and transformer for 3D medical image segmentation. arXiv preprint arXiv:2103.03024 (2021)

Yu, L., et al.: Automatic 3D cardiovascular MR segmentation with densely-connected volumetric ConvNets. In: Descoteaux, M., Maier-Hein, L., Franz, A., Jannin, P., Collins, D.L., Duchesne, S. (eds.) MICCAI 2017. LNCS, vol. 10434, pp. 287–295. Springer, Cham (2017). https://doi.org/10.1007/978-3-319-66185-8_33

Zhang, Y., Liu, H., Hu, Q.: Transfuse: fusing transformers and CNNs for medical image segmentation. arXiv preprint arXiv:2102.08005 (2021)

Zhou, Z., Rahman Siddiquee, M.M., Tajbakhsh, N., Liang, J., et al.: UNet++: a nested U-Net architecture for medical image segmentation. In: Stoyanov, D. (ed.) DLMIA/ML-CDS -2018. LNCS, vol. 11045, pp. 3–11. Springer, Cham (2018). https://doi.org/10.1007/978-3-030-00889-5_1

Attention-Guided Pancreatic Duct Segmentation from Abdominal CT Volumes

Chen Shen[1(✉)], Holger R. Roth[2], Hayashi Yuichiro[1], Masahiro Oda[1], Tadaaki Miyamoto[3], Gen Sato[3], and Kensaku Mori[1]

[1] Graduate School of Informatics, Nagoya University, Nagoya, Japan
cshen@mori.m.is.nagoya-u.ac.jp
[2] NVIDIA Corporation, Santa Clara, USA
[3] Chiba Kensei Hospital, Chiba, Japan

Abstract. Pancreatic duct dilation indicates a high risk of pancreatic ductal adenocarcinoma (PDAC), the deadliest cancer with a poor prognosis. Segmentation of dilated pancreatic duct from CT taken from patients without PDAC shows the potential to assist the early detection of PDAC. Most current researches include pancreatic duct segmentation as one additional class for patients who have already detected PDAC. However, the dilated pancreatic duct for people who have not yet developed PDAC is typically much smaller, making the segmentation difficult. Deep learning-based segmentation on tiny components is challenging because of the large imbalance between the target object and irrelevant regions. In this work, we explore an attention-guided approach for dilated pancreatic duct segmentation as a screening tool for pre-PDAC patients, enhancing the pancreas regions' concentration and ignoring the unnecessary features. We employ a multi-scale aggregation to combine the information at different scales to improve the segmentation performance further. Our proposed multi-scale pancreatic attention-guided approach achieved a Dice score of 54.16% on dilated pancreatic duct dataset, which shows a significant improvement over prior techniques.

Keywords: Pancreatic duct dilation · Pancreatic duct segmentation · Attention

1 Introduction

Pancreatic cancer is one of the most intractable cancers with the lowest 5-year relative survival rate of approximately 10% in the USA. The most common pancreatic cancer occurs in the main pancreatic duct, known as pancreatic ductal adenocarcinoma (PDAC) [11]. Pancreatic duct dilation is identified as a high risk of PDAC in several clinical studies [3,20]. A comparison of pancreatic duct dilation and normal pancreas is shown in Fig. 1. Because of the pancreatic duct's

C. Oyarzun Laura et al. (Eds.): CLIP/DCL/LL-COVID/PPML 2021, LNCS 12969, pp. 46–55, 2021.
https://doi.org/10.1007/978-3-030-90874-4_5

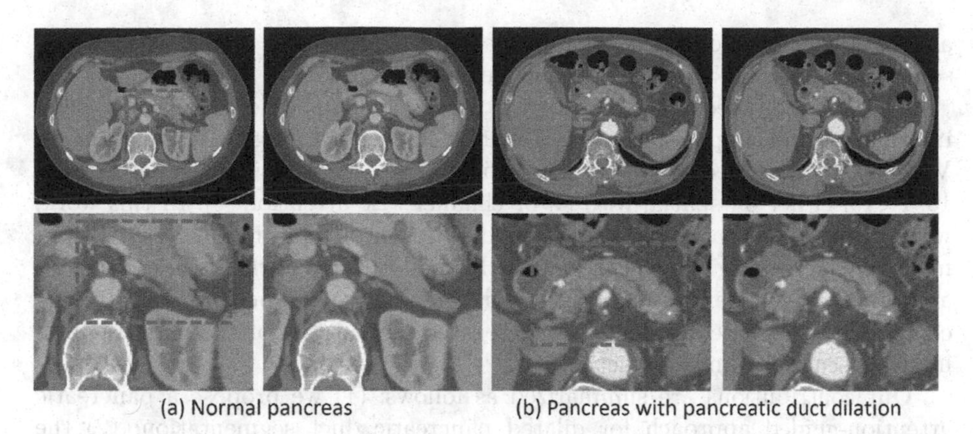

Fig. 1. A comparison of (a) normal pancreas (yellow) and (b) pancreas with dilated pancreatic duct (red). For normal pancreas, the pancreatic duct is invisible on CT. Blue boundaries indicate the pancreas regions. (Color figure online)

slenderness, the normal pancreatic duct region is invisible on contrast-enhanced abdominal CT. Visibility of pancreatic duct from CT modality even could be a direct warning sign for PDAC. Hence, automated dilated pancreatic duct segmentation from CT is promising for the early diagnosis of PDAC as a clinical screening tool.

Only few studies are related to pancreatic duct segmentation [18,21,23]. Zhou et al. [23] investigated a dual-path network for pancreas, PDAC tumor, and pancreatic duct segmentation. Both arterial and venous phase CT volumes are required to conduct the high performance of segmentation performance. Xia et al. [21] proposed a segmentation method based on multi-phase CT alignment that outperforms Zhou et al.'s performance on same dataset. However, both of these researches are focused on PDAC patients, which means that the pancreatic duct region is much thicker than the people who have not developed PDAC yet. Moreover, they both require manual annotations of the pancreatic duct on multiple-phase CT volumes, which is expensive and time consuming.

To the best of our knowledge, only one existing work is concerning the dilated pancreatic duct on the pre-PDAC patients [18]. This work introduced a cascade framework to dilated pancreatic duct segmentation. They generate the region of interest (ROI) of pancreas regions by utilizing the public TCIA pancreas dataset [13], and then make the dense prediction of pancreatic duct on the ROI crops. The difficulties of pancreatic duct segmentation are mainly caused by the small size of duct regions. Additionally, the intensity similarity usually leads to the false-positive prediction outside the pancreas regions.

To address these challenges, we proposed a pancreatic attention-guided approach that learns to focus on pancreas regions during the pancreatic duct segmentation. The general experience motivating our approach is that when radiologists look for the pancreatic ducts, they only concentrate on the pancreas regions and ignore other regions' contexts. This suggests that a deep neural network should also only concentrate on the organ that includes the target object when we do some specific task like pancreatic duct segmentation.

The attention mechanism is widely utilized to enhance the influence of useful information and suppress the useless context in deep learning-based approaches. Various attention techniques are investigated for medical image analysis [12, 15–17,19] either on spatial-wise [4] or channel-wise attention [6]. In this work, we propose an attention-guided method by using an automatically predicted mask of pancreas regions to enhance the learned features of the pancreas regions and ignore the irrelevant regions. To further improve the performance on tiny component of pancreatic duct, we aggregates the outputs of the network from multiple scales for final prediction.

Our contributions are summarized as follows: (1) we propose a pancreatic attention-guided approach for dilated pancreatic duct segmentation; (2) the multi-scale aggregation is adopted to make full use of the information generated in different scales; (3) we validate our approach on a real-world dilated pancreatic duct dataset from patients who have not detected as PDAC yet and achieve state-of-the-art performance on pancreatic duct segmentation using single phase CT volumes only.

2 Methods

With the increasing power of deep learning, convolutional neural networks (CNNs) methods have effectively solved many challenging tasks, such as classification, segmentation or object detection. Fully convolutional network (FCNs) [8] showed great successes in medical image analysis, e.g., on organ segmentation[5], lesion detection [22] tasks.

Our aim is to investigate dilated pancreatic duct segmentation method using the predicted mask of the pancreas. We employed an FCN architecture based on 3D U-Net illustrated in Fig. 2. We have a training set $\boldsymbol{S} = \{(\boldsymbol{I}_n, \boldsymbol{L}_n, \boldsymbol{P}_n), n = 1, ..., N\}$, where $\boldsymbol{I}_n \in \mathcal{R}^{W_n \times H_n \times D_n}$ is the n-th volume of the total N CT volumes with the size of $W_n \times H_n \times D_n$ in width, height and depth, respectively. $\boldsymbol{L}_n$ is the corresponding voxel-wise ground truth volumes of the dilated pancreatic duct, and $\boldsymbol{P}_n$ is the corresponding coarse pancreas prediction, which gained by utilizing TCIA pancreas dataset [13], followed by previous work [18]. Our aim is to learn a model to predict the dilated pancreatic duct $\hat{\boldsymbol{L}} = f(\boldsymbol{I}, \boldsymbol{P})$ with the use of CT volumes and the corresponding coarse pancreas predictions.

2.1 Pancreatic Attention-Guide

Attention mechanism is widely utilized in medial image analysis field [12,15,16, 19], to capture the useful information and ignore the useless context in FCNs. To effectively focus on the organ that the target object belongs to, we proposed a pancreatic attention-guided method inspired by the grid-attention method [12]. Instead of using the image spatial information obtained in the bottleneck, we

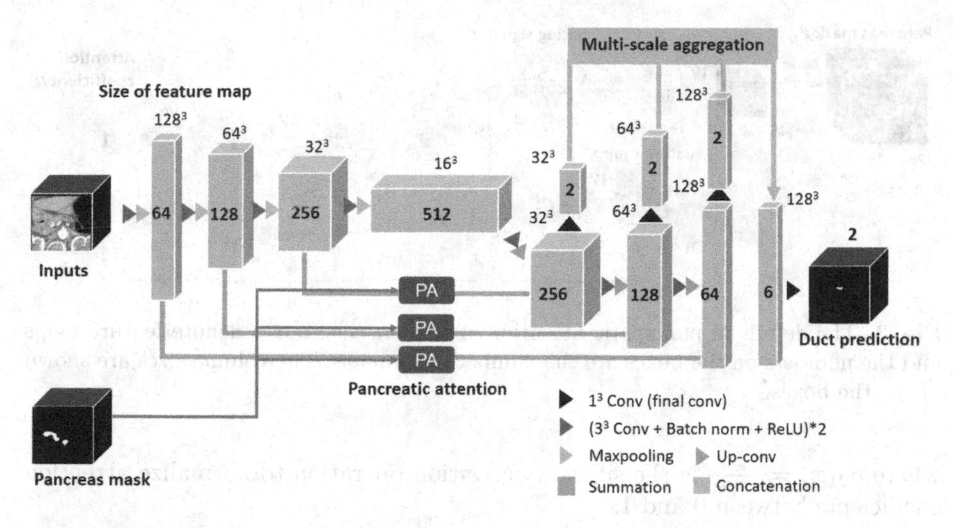

Fig. 2. The architecture of proposed multi-scale pancreatic attention-guided network. The boxes denote feature maps, and the numbers on the boxes are the number of channels. The volume size is above the boxes. *PA* in blue boxes indicates the pancreatic attention operation, which is described in detail in Fig. 3 (Color figure online)

introduce pancreatic attention, which fully exploited the pancreas region's coarse prediction.

The details of pancreatic attention is shown in Fig. 3. The attention coefficients $\alpha_n^l \in [0, 1]$ of n-th volume is calculated at each level l to re-scale the feature map $\boldsymbol{x}_n^l$ by the element-wise multiplication: $\hat{\boldsymbol{x}}_n^l = \alpha_n^l \cdot \boldsymbol{x}_n^l$. The coarse pancreas prediction is used as a mask to guide the grid-attention. An adaptive averaging pooling [9] $\mathcal{Z}(\boldsymbol{P}_n)$ is used to downsample the pancreas mask to meet the size of bottleneck layer. A linear transformation with $1 \times 1 \times 1$ convolution $\boldsymbol{W}_g$ is used as the gating function to learn the pixel-wise focus regions from the pancreas mask, and the corresponding bias is b_g. The gating signal vector $\boldsymbol{\mathcal{G}}_n$ is formulated as

$$\boldsymbol{\mathcal{G}}_n = \phi(\boldsymbol{W}_g \mathcal{Z}(\boldsymbol{P}_n) + b_g), \tag{1}$$

which used to calculate the focus regions. Where ϕ indicates the trilinear interpolation to upsample signal to match up the size to the feature map. For the input feature map, a $2 \times 2 \times 2$ convolution with stride 2 is employed as $\boldsymbol{W}_\theta$. An additive attention $\boldsymbol{q}_n^l$ can be obtained by channel-wise addition , formulated as

$$\boldsymbol{q}_n^l = \sigma_1(\boldsymbol{W}_\theta \boldsymbol{x}_n^l + \boldsymbol{\mathcal{G}}_n), \tag{2}$$

where $\sigma_1(\boldsymbol{x}) = max(0, \boldsymbol{x})$ is the rectified linear unit activation operation. To get the output of attention coefficient, a $1 \times 1 \times 1$ convolution with stride 1 is used to compute the final output $\boldsymbol{W}_\psi$, and b_ψ is the corresponding bias. The formulation of attention coefficients α_n^l is formulated as

$$\alpha_n^l = \phi(\sigma_2(\boldsymbol{W}_\psi \boldsymbol{q}_n^l + b_\psi)) \tag{3}$$

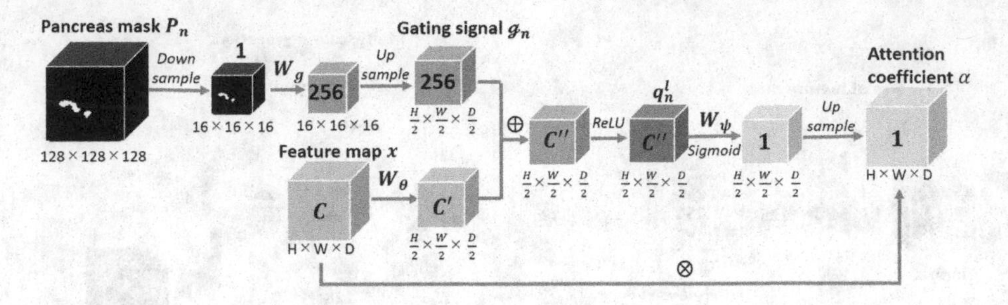

Fig. 3. The details of pancreatic attention operation. The boxes denote feature maps and the numbers on the boxes are the number of channels. The volume sizes are shown below the boxes.

where $\sigma_2(\boldsymbol{x}) = \frac{e^{\boldsymbol{x}}}{e^{\boldsymbol{x}}+1}$ is the sigmoid activation operation to normalize attention coefficients between 0 and 1.

2.2 Multi-scale Aggregation

Features from different scales have different influences on the final prediction. Features from low resolution focus on the semantic information, and the high resolution is concerning more on spatial information. We aggregate the attention vector from each level for final segmentation. Aggregation of feature proven to have amazing performance in segmentation [1,14,19]. We follow the attention aggregation introduced in [16]. Feature on each scale is indicated as $\boldsymbol{F}_j$, where $\{j = 1, 2, 3\}$ indicates the j-th level of the network, except the bottleneck layer. The final prediction of each scale j can be obtained by using an $1 \times 1 \times 1$ convolution to transform the number of channels to the number of the output classes K. Due to the resolution difference among feature maps at each scale, the bilinear interpolations are used to upsample the feature map to $\hat{\boldsymbol{F}}_j$. The output $\mathcal{F}$ of the network is formulated as

$$\mathcal{F} = \boldsymbol{W}_f([\hat{\boldsymbol{F}}_1, \hat{\boldsymbol{F}}_2, \hat{\boldsymbol{F}}_3]), \tag{4}$$

where $\boldsymbol{W}_f$ is an $1 \times 1 \times 1$ final convolution to reduce the the channels of concatenated feature map to class number K.

3 Experiments and Results

3.1 Dataset and Settings

Dataset. We evaluated our proposed methods on a dilated pancreatic duct dataset reported in [18]. It consists of 30 contrast-enhanced abdominal CT volumes collected from people with normal pancreas (i.e. pre-PDAC) but pancreatic duct dilation. Each CT volume contains 192-887 slices with size 512 × 512 pixels, and the resolutions of CT volumes are [0.59–0.75, 0.59–0.75, 0.3–1.0] mm. The annotations of pancreatic duct regions are manually generated by semi-supervised annotation tools using region growing algorithm. The pancreas coarse segmentation mask and ROI crops of pancreas are gained following [18]. The dataset is divided into five equal folds, including 6 CT volumes for each fold. One fold is fixed as the testing set, and the other 4 folds were adopted for cross-validation throughout the experiments.

Implementation. Our experiments were implemented with PyTorch 1.6.0. For data pre-processing, we clip the intensity of all CT volumes into the range of [−200, 200] HU and adapted min-max normalization to rescale the values into the range of [0, 1]. The resolution CT volumes were resampled into 1 mm isotropic. The input and output size of the networks were $128 \times 128 \times 128$ voxel, and the mini-batch size was two. The Adam optimization [7] is employed with the initial learning rate of 10^{-4} to minimize the Dice loss function proposed in [10]. No additional data augmentation is used in our experiments. All experiments are performed on an NVIDIA Quadro P6000 with 24 GB memory.

Evaluation. We employ Dice score (DSC) and sensitivity as measures to evaluate our segmentation results. $DSC = \frac{2TP}{2TP+FP+FN}$ measures the overlap ratio between ground truth and segmentation result, where voxel numbers of true positive, false positive and false negative are indicated as TP, FP and FN, respectively. The range of DSC is [0,1], which 1 mean indicates the perfect segmentation. $Sensitivity = \frac{TP}{TP+FN}$ measures the portion of TP.

3.2 Segmentation Results and Discussion

To show the effectiveness of the attention-guided approach as well as the multi-scale aggregation on dilated pancreatic duct segmentation, we first compared the accuracies of segmentation results using 3D U-Net [2], SE-Dense U-Net [18], Attention U-Net [12], and our proposed pancreatic-attention guided network (PA Net) as well as multi-scale PA Net (MPA Net) in Table 1. Note that the same dataset is used in this comparison. DSC and sensitivity of our proposed methods are 54.16% and 61.70%, which significantly outperform the state-of-the-art performance 49.87% and 51.94% reported in [18]. The ablation study of multi-scale aggregation shows the advantages of applying information from different resolutions in FCNs. We have to mention that our proposed methods can segment pancreatic duct successfully in all testing cases, compared the minimum DSC and sensitivity of 0% in other methods. Figure 4 shows segmentation examples of axial slice and 3D rendering in each method.

The performance of pancreatic attention-guided we introduced beat the original Attention U-Net in this work. This might be due to the fact that original attention U-Net uses the context of the bottleneck, which might not suit for relatively small objects like the pancreatic duct. Because pancreatic ducts are the extremely tiny component that only occupy a small part of the image, learning the attention gating function directly is challenging and some important training information might be missed. This assumption is supported by the visualization

Table 1. Comparison of Dice score and sensitivity for dilated pancreatic duct segmentation with 3D U-Net [2], SE-Dense U-Net [18], Attention U-Net [12], PA Net, and MPA Net. Best scores are shown in **bold**.

Methods	Dice (%)			Sensitivity (%)		
	Mean	Min	Max	Mean	Min	Max
3D U-Net [2]	47.07%	0.00%	**78.26%**	50.54%	0.00%	87.75%
SE-Dense U-Net [18]	49.87%	0.00%	74.11%	51.94%	0.00%	92.50%
Attention U-Net [12]	44.73%	0.00%	75.49%	52.49%	0.00%	86.25%
PA Net (proposed)	53.20%	**21.41%**	74.21%	61.63%	**36.28%**	89.75%
MPA Net (proposed)	**54.16%**	15.81%	75.05%	**61.70%**	22.28%	**94.50%**

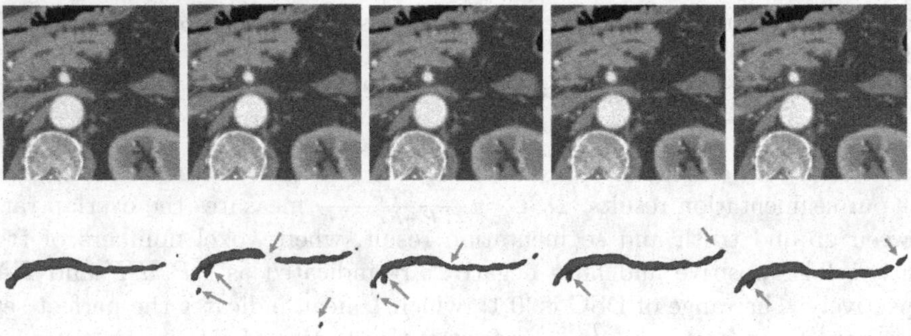

(a) GT (b) SED U-Net (c) Att U-Net (d) PA Net (e) MPA Net

Fig. 4. Comparison of dilated pancreatic duct segmentation results in axial slice and 3D rendering on (a) ground truth (GT), (b) SE-Dense U-Net (SED U-Net), (c) Attention U-Net (Att U-Net) [12] and our proposed (d) PA Net and (e) MPA Net. The pancreatic duct regions are shown in red. Blue arrows point to the regions are not segmented well. (Color figure online)

of feature attention coefficients shown in Fig. 5. Some pancreatic duct parts were not included in the focus when using Attention U-Net. That is likely why the original Attention U-Net performed even worse than the standard 3D U-Net. Our proposed PA Net and MPA Net effectively focused on the whole pancreas regions that are known to include the pancreatic duct.

The comparison of DSC on pancreatic duct segmentation with other methods is summarized in Table 2. Note that both of the results reported in [23] and [21] were reported on pancreatic ducts from PDAC patients, who have much larger pancreatic ducts. Also, a larger dataset with 239 cases is used compared to our 30 cases. While direct comparison is not possible, our methods achieve the highest reported DSC on pancreatic duct segmentation on single-phase CT volumes using a smaller dataset. Our performance is however no rival for methods using multiple phases CT volumes. Multi-phase CT volumes assist in learning more

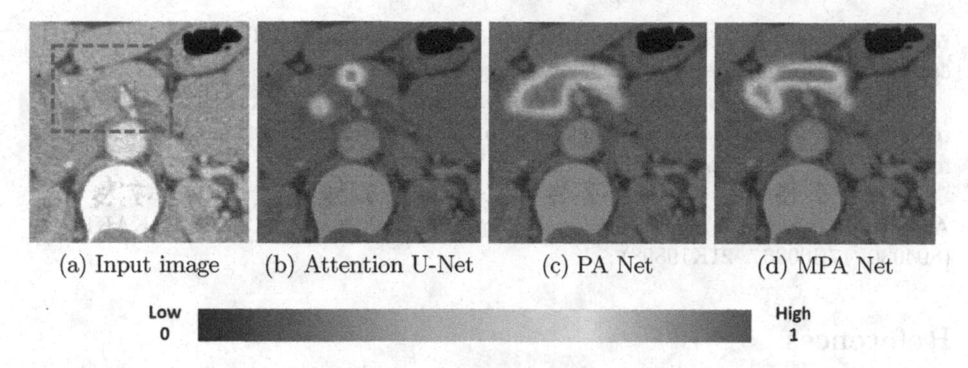

Fig. 5. Visualization of feature attention coefficients on (a) input image by using (b) Attention U-Net [12] and our proposed (c) PA Net and (d) MPA Net. Blue boundary in the input image indicates the pancreas region. (Color figure online)

Table 2. Comparison to other pancreatic duct segmentation methods. Here we listed other segmentation results performed on single/multiple phases CT volumes. Our proposed methods are shown in *Italic*.

Methods	Phase	# of data	Dice (pancreatic duct)
3D-UNet-single-phase (Arterial) [23]	Single phase	239 PDAC	38.35 ± 28.98
3D-UNet-single-phase (Venous) [23]	Single phase	239 PDAC	40.25 ± 27.89
3D-ResDSN-single-phase (Arterial)[23]	Single phase	239 PDAC	47.04 ± 26.42
3D-ResDSN-single-phase (Venous) [23]	Single phase	239 PDAC	49.81 ± 26.23
Cascade SE-Dense U-net [18]	Single phase	30 normal	49.87 ± 22.54
PA Net (proposed)	*Single phase*	*30 normal*	*53.20 ± 11.52*
MPA Net (proposed)	*Single phase*	*30 normal*	*54.16 ± 12.60*
3D-UNet-multi-phase (fusion) [23]	Multi phase	239 PDAC	39.06 ± 27.33
3D-UNet-multi-phase-HPN [23]	Multi phase	239 PDAC	44.93 ± 24.88
3D-ResDSN-multi-phase (fusion) [23]	Multi phase	239 PDAC	48.49 ± 26.37
3D-ResDSN-multi-HPN [23]	Multi phase	239 PDAC	56.77 ± 23.33
Multi-phase Alignment Ensemble [21]	Multi phase	239 PDAC	64.38 ± 29.67

information during training and inference. But, the annotated pancreatic duct dataset from multiple phases is extremely hard to obtain.

4 Conclusion

In this paper, we proposed a pancreatic attention-guide method for dilated pancreatic duct segmentation. Our approach was inspired by clinical experience: when radiologists go through the CT volumes for pancreatic duct dilation, they naturally neglect the regions outside of the pancreas. We employed the coarse pancreas prediction result to assist the learning of the attention coefficients, which eliminates the influence outside the pancreas regions. Moreover, we included multi-scale aggregation to make full use of the information learned on diverse feature resolutions. No basic data augmentation was utilized in this work. The average DSC and sensitivity are 54.16% and 61.70%, which dramatically outperforms the prior method using the same dataset.

Our methods could apply to other medical imaging tasks where the organ-relationship is know and can act as a prior to guide subsequent segmentation models, such as organ-specific tumors.

Acknowledgement. This work was supported by the MEXT/JSPS KAKENHI (894030, 17H00867, 21K19898).

References

1. Chen, L., Yang, Y., Wang, J., Xu, W., Yuille, A.L.: Attention to scale: scale-aware semantic image segmentation. In: 2016 IEEE Conference on Computer Vision and Pattern Recognition (CVPR), pp. 3640–3649 (2016). https://doi.org/10.1109/CVPR.2016.396
2. Çiçek, Ö., Abdulkadir, A., Lienkamp, S.S., Brox, T., Ronneberger, O.: 3D U-net: learning dense volumetric segmentation from sparse annotation. In: Ourselin, S., Joskowicz, L., Sabuncu, M.R., Unal, G., Wells, W. (eds.) MICCAI 2016. LNCS, vol. 9901, pp. 424–432. Springer, Cham (2016). https://doi.org/10.1007/978-3-319-46723-8_49
3. Edge, M.D., Hoteit, M., Patel, A.P., Wang, X., Baumgarten, D.A., Cai, Q.: Clinical significance of main pancreatic duct dilation on computed tomography: single and double duct dilation. World J. Gastroenterol.: WJG **13**(11), 1701 (2007)
4. Hu, J., Shen, L., Sun, G.: Squeeze-and-excitation networks. In: Proceedings of the IEEE Conference on Computer Vision and Pattern Recognition, pp. 7132–7141 (2018)
5. Isensee, F., Kickingereder, P., Wick, W., Bendszus, M., Maier-Hein, K.H.: No new-net. In: Crimi, A., Bakas, S., Kuijf, H., Keyvan, F., Reyes, M., van Walsum, T. (eds.) BrainLes 2018. LNCS, vol. 11384, pp. 234–244. Springer, Cham (2019). https://doi.org/10.1007/978-3-030-11726-9_21
6. Khandelwal, S., Sigal, L.: AttentionRNN: a structured spatial attention mechanism. In: Proceedings of the IEEE/CVF International Conference on Computer Vision (ICCV), October 2019
7. Kingma, D.P., Ba, J.: Adam: a method for stochastic optimization. In: International Conference for Learning Representations (2015)
8. Long, J., Shelhamer, E., Darrell, T.: Fully convolutional networks for semantic segmentation. In: Proceedings of the IEEE Conference on Computer Vision and Pattern Recognition, pp. 3431–3440 (2015)
9. McFee, B., Salamon, J., Bello, J.P.: Adaptive pooling operators for weakly labeled sound event detection. IEEE/ACM Trans. Audio Speech Lang. Process. **26**(11), 2180–2193 (2018). https://doi.org/10.1109/TASLP.2018.2858559
10. Milletari, F., Navab, N., Ahmadi, S.: V-Net: fully convolutional neural networks for volumetric medical image segmentation. In: 2016 Fourth International Conference on 3D Vision (3DV), pp. 565–571 (2016). https://doi.org/10.1109/3DV.2016.79
11. Mizrahi, J.D., Surana, R., Valle, J.W., Shroff, R.T.: Pancreatic cancer. Lancet **395**(10242), 2008–2020 (2020)
12. Oktay, O., et al.: Attention U-net: learning where to look for the pancreas. arXiv preprint arXiv:1804.03999 (2018)

13. Roth, H.R., Farag, A., Turkbey, E.B., Lu, L., Liu, J., Summers, R.M.: Data from pancreas-CT (2016). https://doi.org/10.7937/K9/TCIA.2016.tNB1kqBU
14. Roth, H.R., et al.: A multi-scale pyramid of 3D fully convolutional networks for abdominal multi-organ segmentation. In: Frangi, A.F., Schnabel, J.A., Davatzikos, C., Alberola-López, C., Fichtinger, G. (eds.) MICCAI 2018. LNCS, vol. 11073, pp. 417–425. Springer, Cham (2018). https://doi.org/10.1007/978-3-030-00937-3_48
15. Roy, A.G., Navab, N., Wachinger, C.: Concurrent spatial and channel 'squeeze & excitation' in fully convolutional networks. In: Frangi, A.F., Schnabel, J.A., Davatzikos, C., Alberola-López, C., Fichtinger, G. (eds.) MICCAI 2018. LNCS, vol. 11070, pp. 421–429. Springer, Cham (2018). https://doi.org/10.1007/978-3-030-00928-1_48
16. Schlemper, J., et al.: Attention-gated networks for improving ultrasound scan plane detection. arXiv preprint arXiv:1804.05338 (2018)
17. Schlemper, J., et al.: Attention gated networks: learning to leverage salient regions in medical images. Med. Image Anal. **53**, 197–207 (2019)
18. Shen, C., et al.: A cascaded fully convolutional network framework for dilated pancreatic duct segmentation. In: 35th International Congress and Exhibition on Computer Assisted Radiology (2021)
19. Sinha, A., Dolz, J.: Multi-scale self-guided attention for medical image segmentation. IEEE J. Biomed. Health Inform. **25**(1), 121–130 (2021). https://doi.org/10.1109/JBHI.2020.2986926
20. Tanaka, S., et al.: Main pancreatic duct dilatation: a sign of high risk for pancreatic cancer. Jpn. J. Clin. Oncol. **32**(10), 407–411 (2002)
21. Xia, Y., et al.: Detecting pancreatic ductal adenocarcinoma in multi-phase CT scans via alignment ensemble. In: Martel, A.L. (ed.) MICCAI 2020. LNCS, vol. 12263, pp. 285–295. Springer, Cham (2020). https://doi.org/10.1007/978-3-030-59716-0_28
22. Yan, K., Wang, X., Lu, L., Summers, R.M.: DeepLesion: automated mining of large-scale lesion annotations and universal lesion detection with deep learning. J. Med. Imaging **5**(3), 1–11 (2018). https://doi.org/10.1117/1.JMI.5.3.036501
23. Zhou, Y., et al.: Hyper-pairing network for multi-phase pancreatic ductal adenocarcinoma segmentation. In: Shen, D., et al. (eds.) MICCAI 2019. LNCS, vol. 11765, pp. 155–163. Springer, Cham (2019). https://doi.org/10.1007/978-3-030-32245-8_18

Development of the Next Generation Hand-Held Doppler with Waveform Phasicity Predictive Capabilities Using Deep Learning

Adrit Rao[1(✉)], Akshay Chaudhari[2], and Oliver Aalami[2]

[1] Palo Alto High School, Palo Alto, CA, USA
[2] Stanford University, Stanford, CA, USA
{akshaysc,aalami}@stanford.edu

Abstract. The most ubiquitous tool utilized at the point-of-care within clinic for the evaluation of Peripheral Arterial Disease (PAD) or limb perfusion is the hand-held doppler. Providers interpret audible circulatory sounds in order to predict waveform phasicity as Monophasic (severe disease), Biphasic (moderate disease), or Triphasic (normal). This prediction is highly subjective. With any doubt in interpretation, patients are referred to an ultrasound laboratory for formal evaluation which can delay definitive diagnosis and treatment. This paper proposes a novel deep learning system integrated within a hardware doppler platform for low-cost, fast, and accessible assessment of disease at the point-of-care. The system converts input sound into a spectrogram and classifies waveform phasicity with a neural network through learned visual temporal amplitude patterns. The hardware system consists of a low-cost single board computer and simple peripherals including a microphone and LCD display. The final system received a validation accuracy of 96.23%, an F1 score of 90.57%, and was additionally tested within a manually simulated clinical environment with interference in which it received an accuracy of 91.66%. Such a tool can be used to more objectively and efficiently diagnose PAD at the point-of-care and prevent delays in intervention.

Keywords: Hand-held Doppler · Arterial disease · Deep learning

1 Introduction

1.1 Background

Peripheral Arterial Disease (PAD) is the narrowing or blockage of arteries supplying the lower arterial extremities due to atherosclerosis or the build-up of plaque [7–9,15]. According to the Center for Disease Control (CDC) over 6.5 million people who are 40 years or older have PAD in the United States alone [2]. If not diagnosed and managed at an early stage in disease progression, PAD can lead to amputation causing life-long disability [6]. After clinical examination

C. Oyarzun Laura et al. (Eds.): CLIP/DCL/LL-COVID/PPML 2021, LNCS 12969, pp. 56–67, 2021.
https://doi.org/10.1007/978-3-030-90874-4_6

for a pedal pulse, the most common point-of-care tool utilized for the evaluation of PAD is a low-cost hand-held continuous wave doppler [22]. The hand-held doppler enables practitioners to "listen" to arterial circulatory sound through ultrasound waves. This sound is manually assessed and categorized into a waveform phasicity class. Monophasic waveforms represent severe arterial insufficiency, Biphasic waveforms represent moderate arterial insufficiency, and Triphasic waveforms are considered normal [18]. Through the interpretation of sound characteristics, practitioners predict phasicity, which is a technically challenging and subjective procedure. With any doubt in interpretation, patients will be referred to receive further assessment within a formal vascular ultrasound laboratory for the measurement of the Ankle-Brachial Index (ABI) [4]. This leads to long delays in objective diagnosis that further delays more definitive treatment, such as placing compression wraps for edema or referring patients to a vascular specialist. The ability to accurately derive phasicity at the point-of-care ubiquitously can enable timely and robust diagnosis and treatment of PAD.

Deep learning is the ability by which artificial neural structures can learn and make predictions from data similar to the human brain [12]. The capability of developing artificial neural networks (ANNs) is starting to be widely leveraged in the medical domain to aid clinicians in the diagnosis of disease [13]. Current medical applications of deep learning are more focused on medical imaging tasks using computer vision and are not widely used for the classification of sound. However, the ability to utilize such technology in order to classify arterial sounds from tibial vessels provides the opportunity for accurate, objective, low-cost and accessible aid for providers in the diagnosis of disease at the point-of-care. Deep learning techniques have not yet been leveraged to classify sound from tibial vessels in such a manner to derive the waveform phasicity metric for PAD.

1.2 Innovation

In our previous preliminary work [17], we propose a deep learning system for predicting waveform phasicity from arterial circulatory sounds. Our system converts input sound into a spectrogram enabling visual differentiation between the phasicity classes based off of temporal patterns. Through the conversion of sound into a pixel-by-pixel image representation, we enable the use of a standard deep learning approach for image classification in order to predict phasicity through learned visual features. With our system, we reach a validation accuracy and F1 score of over 90%. In this paper, we integrate our deep learning system within an all-in-one hardware configuration and build robustness and resilience in our system to noisy clinical settings through filtering techniques. With this, we seek to enable a low-cost, accessible and highly accurate tool that can be used directly at the point-of-care in order to enable robust limb perfusion assessments. Additionally, we maximize accessibility by optimizing our computational system to run on the edge removing the reliance on internet or cloud connectivity.

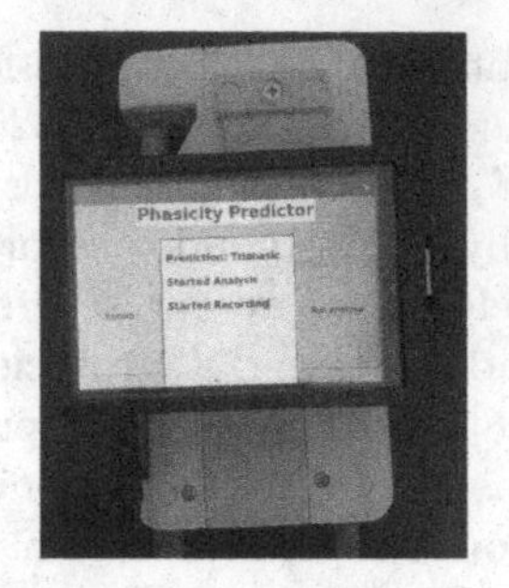

Fig. 1. Deep Learning Doppler Prototype. Left: front view, Right: back view. Consists of touch based LCD display with a custom GUI, microphone and single-board computer within a 3D enclosure. The system is attached to the back of the doppler and can be used while placing the probe on a target artery in a standard evaluation of disease.

1.3 Implementation Summary

Our continuous wave doppler system was developed with a low-cost Raspberry Pi single-board computer processor [16], a microphone and an LCD display. Due to our system not requiring high amounts of compute, we enable the use of a very low-cost hardware platform which in total costs $67. An image taken of our system attached to a standard hand-held doppler (Huntleigh Dopplex MD2) is shown in Fig. 1. The doppler consists of a pencil probe that is placed on a target artery and a speaker which relays the circulatory sounds using doppler ultrasound waves. Our prototype is positioned onto the back of the doppler and uses a microphone in order to record output sound. Our custom Graphical User Interface (GUI) is shown on the touch-based LCD display and consists of a simple view used for recording and performing deep learning analysis. Between the two buttons, we use a console-based text view in order to inform the user on actions such as starting and ending a recording along with performing analysis.

2 Methods

To develop our software and hardware system, we collected doppler audio data as part of a study within a formal ultrasound laboratory in patients being assessed for vascular disease (PAD). We then generated a spectrogram image dataset from the sound files as input to our deep learning model. We used a custom filtering technique in order to gain system resilience to background noise factors. Next, we construct and train a convolutional neural network (CNN) with an Inception V3 [21] backbone and employed a transfer learning approach for re-training. Lastly, we constructed our hardware platform and integrated our software system.

Our end-to-end computational system consisting of spectrogram conversion along with CNN input and prediction is shown as a visual flow-chart in Fig. 2.

The following method sections covers implementation details starting with data preparation consisting of collection, generation, filtering and augmentation.

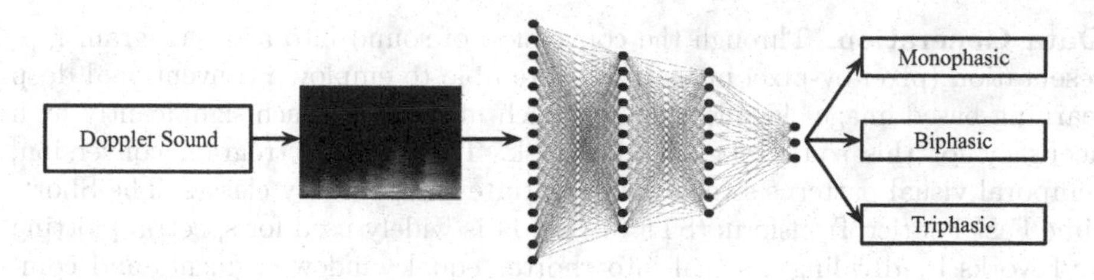

Fig. 2. Waveform Phasicity Prediction Deep Learning System Architecture. Input doppler sound is converted into a spectrogram and a CNN-based image classifier uses learned visual features to categorize the circulatory sound into a phasicity class.

Next is model development which covers the standard approach used in order to train the Inception V3 CNN architecture on the spectrogram dataset and small modifications used to accommodate our dataset. After this, we cover the main innovation of this paper, the hardware platform consisting of our hardware construction and software integration. Our full system was implemented in Python v3.7 in the Google Colab Pro Integrated Development Environment (IDE) [5] with an NVIDIA Tesla V-100 Graphics Processing Unit (GPU) [1]. For validation purposes, we used Matplotlib [10] for embedded graphing of accuracy/loss.

2.1 Data Preparation

Data Collection. In this IRB-approved study, patients referred to be evaluated for PAD were asked to participate. If agreeable, consent was obtained and data was collected by a vascular technologist performing the Ankle Brachial Index (ABI) study. We developed a custom iOS application for the collection of doppler audio data (Swift 5, Xcode IDE). The app allowed vascular technologists to record and enter waveform phasicity, Ankle-Brachial Index (ABI), artery name, and laterality in patients being assessed for PAD and other cardiovascular diseases within an IAC accredited vascular ultrasound laboratory. The Parks Flo-Lab 2100 doppler ultrasound machine (Parks Medical Electronics, Inc., Aloha, OR) was used as a robust reference in order to derive ground-truth labeling and receive high-quality audio samples along with keeping data consistency. To maintain a structured dataset, recordings were capped at 4 s in length. For model training, 80% of the data was used and 20% was used for baseline validation. In total, we collected 268 audio recording samples of which we had 102 Triphasic (38.1%), 80 Biphasic (29.9%), and 86 Monophasic (32.1%) recordings. Collected data spanned 67 patients with diverse vascular conditions with both compressible and non-compressible tibial vessels. After data collection, data was cross-checked by 2 professional vascular clinicians to ensure data integrity and fix incorrectly labeled audio files. Sound files with background noise and interference were intentionally kept in the final dataset with the goal of the model building resilience to clinical noise factors and sound from other machinery.

Data Generation. Through the conversion of sound into a spectrogram representation (pixel-by-pixel image), we were able to employ a conventional deep learning-based image classification approach in order to reach significantly high accuracy for this sound classification task. Through spectrogram conversion, temporal visual patterns strongly differentiate the phasicity classes. The Short-time Fast Fourier Transform (STFT) (Eq. 1) is widely used for spectral plotting and works by dividing a signal into shorter equal window segments and computing the standard fourier transform. The changes are then plotted as a visual spectrum that represents changes in frequency and amplitude. When assessing arterial sounds within a formal vascular laboratory, the ultrasound machinery plots a waveform with similar differentiating characteristics as the spectrogram images (waveform peaks). Generation was done with the Librosa [14] library.

$$\mathbf{STFT}\{x[n]\}(m,\omega) \equiv X(m,\omega) = \sum_{n=-\infty}^{\infty} x[n]w[n-m]e^{-j\omega m} \tag{1}$$

In this formula, the signal is represented as $x[n]$ and the window as $w[n]$.

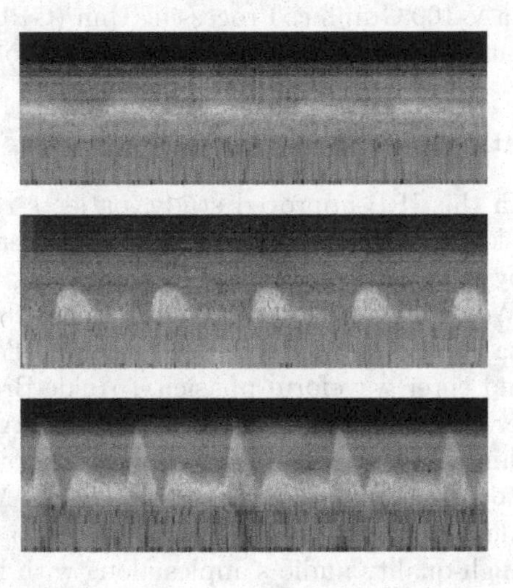

Fig. 3. Waveform Phasicity Spectrogram Example Images. Differentiation present through temporal patterns and amplitude peaks. Top: Monophasic (single peak, severe), Mid: Biphasic (two peaks, moderate), Bottom: Triphasic (three peaks, normal).

Examples of generated spectrogram images are shown in Fig. 3. Notice visual temporal differentiation through the number of peaks in each phasicity class. The Monophasic class has low amplitude and a single peak, the Biphasic class has a higher amplitude consisting of two peaks and the Triphasic class consists of three peaks and the highest amplitude. This representation of input doppler circulatory sound enables clear visual differentiation between the phasicity classes.

Data Cleaning. When integrating our system within a hardware platform with the goal of being used at the point-of-care, it is important to take into consideration background noise caused by other machines and sounds present within a clinical setting. When background noise is present, spectrogram images gain unnecessary features which can possibly throw off the model while extracting features and making predictions. To prevent this from happening, we use the Butterworth high pass filter (Eq. 2) [19] in order to filter the spectrogram images. Through performing this signal filter, the upper half features are cleared and the lower half important amplitude peaks remain. This builds robustness in our system to interference within noisy clinical environments due to surrounding sounds. The Butterworth filter was implemented with the Scipy [24] library.

$$G(\omega) = \frac{1}{\sqrt{1+\omega^{2n}}} \tag{2}$$

In this formula, the angular frequency is represented as w (radians per second) and n is the number of poles in the filter. This is the original formulation.

Data Augmentation. As the dataset was fairly small in size, we decided to use artificial data augmentations in order to increase possible learned features from the model in later stages. We solely added 20% of randomized zoom and decided not to use standard data augmentations. The reason for this was because the end system would be receiving clean spectral images that would be in the same orientation. Thus, adding the standard augmentation pipeline consisting of sheer, flips and rotation may add representational capacity for the model that may not be useful at test time. When adding randomized zoom of 20%, amplitude peaks remain in frame and only small percentage of the border is cropped leaving peak features in view while still causing a slight amount of healthy dataset diversity.

2.2 Model Development

In total, the dataset size was likely not large enough in size to train a CNN model from scratch and high accuracy would not be achievable. We instead employed a transfer learning approach which allowed for the re-training of an existing model with prior weights. With transfer learning, pretrained model weights can be fine-tuned for a downstream task of interest allowing for higher accuracy rates then training a model from scratch on smaller sized datasets [23]. Inception V3 is a deep CNN model architecture which has received high accuracy rates on the ImageNet [11] benchmark and was chosen as the backbone architecture for our computational system. CNN architectures are used in many medical tasks for their unique ability to extract features through convolution operations. The Inception V3 architecture is mainly focused on using less computational power through state-of-the art dimensionality reduction techniques and has proven to be more efficient in terms of the number of parameters as well as memory and GPU usage. Because of this, Inception V3 was an ideal architecture to be used within our low computational resource system. To accommodate our spectrogram dataset, we slightly modified the Inception V3 architectue by adding an input layer of 244 × 244 before the architecture and adding a flatten and dense layer (FC = 3, softmax) after the architecture for probability class prediction. We trained the model with transfer learning over 10 epochs and calculate accuracy and loss in order to visualize model progression. We utilized the Adam optimizer and sparse categorical-cross entropy loss for training with a batch size of 32 and LR 0.001. Our computational system was developed using Keras for sequential model construction and Tensorflow [3] for training and optimization.

Prior to our more in depth validation and experimentation, we visualize activation heat-maps from our CNN model to ensure that attention is mainly focused around the main differentiating factor, amplitude peaks. Figure 4 shows the input Triphasic spectrogram image (black and white for heat-map visualization purposes) and the output after the activation heat-map visualization. Notice how the heat-map region has main attention over the amplitude peaks. For this visualization, we use the Grad-CAM algorithm (Gradient-weighted Class Activation Mapping) [20] in Keras with our modified Inception V3 network.

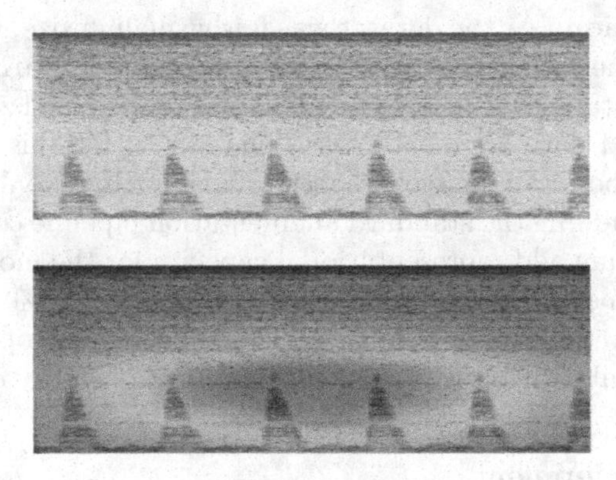

Fig. 4. Grad-CAM Model Activation Visualization (last conv output). Top: Input image, Bottom: Heat-map visualization. Notice attention heat-map over amplitude peaks.

2.3 Hardware Platform

Hardware Construction. We used the Raspberry Pi 3B+ (Raspberry Pi Foundation, Cambridge, UK) single board computer as our hardware compute system. Due to the Raspberry Pi not having an in-built GPU, the use of Inception V3 and its low-computational pipeline potentially helped lower memory usage and processing/inference time. We attach a 3.5 inch 320 × 480 pixel LCD touchscreen display hat (Jun-electron) onto the Raspberry Pi via General Purpose Input/Output (GPIO) pins. We attach a miniature (0.5 inch) USB 2.0 microphone (YOUMI) to the right side of the Raspberry Pi in order to record sound from the doppler. We then enclosed our system within a 3D plastic casing (Jun-electron) and added heat-sinks over the Central Processing Unit (CPU) in order to prevent over-heating. When attaching our integrated system onto the hand-held doppler, we aim to enable ease of use within a fast paced clinical setting where multiple patients are being assessed rapidly. We attached our system to the back of the doppler in order to leave the in-built speaker un-covered as attaching it directly above may cause muffling of the sound output. Additionally, placing it at the back enables easier usage of the touch screen functionality. The standard hand-held doppler has volume control enabling our system to receive clear recordings even when placed on the back of the doppler. We secure our prototype to the doppler using a velcro connector (3M company, Saint Paul, MN) allowing for simple attachment and detachment of the module along with universal usability across almost all standard point-of-care hand-held dopplers.

Software Integration. We developed a custom simplistic GUI optimized to run on the LCD with touch-based actions using Tkinter. The GUI consists of two buttons: one for recording and one for triggering analysis. Between the two buttons, we used a text-view based console in order to inform the user on actions they are taking (started/ended recording) and final predictions (from the CNN model). Our modified Inception V3 CNN model was converted into a Hierarchical Binary Data Format (HDF) in order to load it into our hardware platform. Due to running all computation on-device, no Wi-Fi connectivity is required to run our system. Hardware model inference time ranges between ~10-15 s.

3 Results

We validate our system in two parts, baseline and manual validation. For the baseline validation of our computational system we feed our CNN model 20% of the dataset and calculate the testing accuracy along with the F1 score which is standard for evaluating deep learning-based image classification systems. For manual experimentation, the main goal was to push the limits of our hardware system within a simulated clinical environment to gauge real-world usability. Thus, we construct an experimental situation which consists of playing an industrial machinery sound track in order to generate interference while performing two sub-tests: distance and volume experiments. In the first experiment, we change and increase the distance between the doppler audio output and the device and observe results and in the second experiment we change the volume output of the doppler audio playback and observe results (both experiments use interference sound track playing consistently at 75 dB). Through the use of both a conventional and unconventional validation and experimentation approach, we aim to robustly evaluate our hardware system and understand the limitations.

3.1 Baseline Validation

We use a standard approach which consists of graphing accuracy and loss along with computing the F1 score. Model validation accuracy and loss (Eq. 3) over the 10 training epochs is shown in Fig. 5. As accuracy is consistently increasing and loss is decreasing at a near constant rate, it depicts a good model learning rate. We reached a final validation accuracy of 96.23% and a training accuracy of 90.40%. This proves that the proposed computational system is reasonably accurate in predicting waveform phasicity from doppler sounds through spectrogram conversion and CNN analysis and provides a strong baseline evaluation.

$$Accuracy = \frac{TP + TN}{TP + TN + FP + FN} \tag{3}$$

In this equation, TP represents True Positive, TN represents True Negative, FP represents False Positive and FN represents False Negative results.

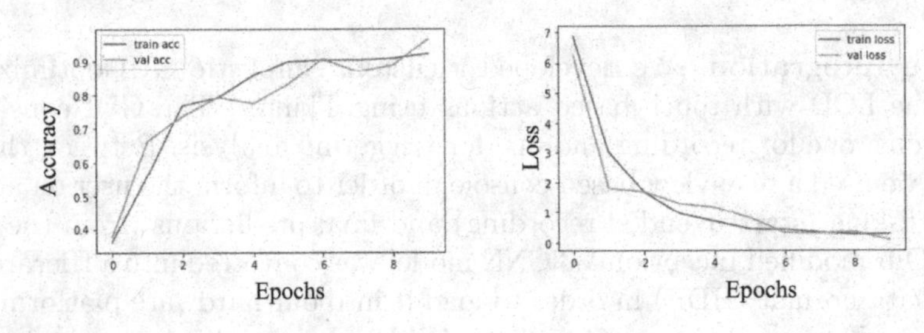

Fig. 5. Accuracy and Loss graphs. 96.23% validation accuracy. 90.40% training accuracy. Consistent accuracy increase and loss decrease.

We additionally calculated the F1 score (Eq. 6), a widely used metric to validate deep learning-based pattern recognition models. The F1 score represents the balance between the precision (Eq. 4) and recall (Eq. 5). After the 10 epochs of training, we receive a proficient F1 score of 90.57%.

$$Precision = \frac{TP}{TP + FP} \tag{4}$$

$$Recall = \frac{TP}{TP + FN} \tag{5}$$

$$F1 = \frac{2 * Precision * Recall}{Precision + Recall} = \frac{2 * TP}{2 * TP + FP + FN} \tag{6}$$

3.2 Manual Experiment

We took an unconventional approach to further validate our hardware system. This experiment consists of playing an Industrial Sound Track in order to simulate a clinical environment to test the efficacy of such a device in a real-world setting. We performed two sub-tests including a distance and volume experiment. In the first test, we changed the distance between the doppler audio playback and the device. In the second test, we changed the output doppler playback volume. These two tests were performed with the goal of pushing the limits of our system and simulating efficacy within a noisy clinical setting with external factors which could effect the performance of the proposed system.

Distance Experiment. In the distance experiment, the doppler playback volume was kept at a constant (85 dB) as we played the industrial sound track. We then validate our hardware system three times for each distance increment (0 in., 3 in., 6 in., 9 in.). The goal of such experimentation is to test the efficacy of the system when the distance between the system and playback is increasing. As the distance increases, the industrial sound track has more of an effect on the spectrogram image putting our spectrogram denoising techniques to the test.

Table 1. Distance experiment results

Distance Level (in)	Prediction	Ground-truth
0 in	Monophasic	Monophasic
0 in	Biphasic	Biphasic
0 in	Triphasic	Triphasic
3 in	Monophasic	Monophasic
3 in	Biphasic	Biphasic
3 in	Triphasic	Triphasic
6 in	Monophasic	Monophasic
6 in	Biphasic	Biphasic
6 in	Triphasic	Triphasic
9 in	Biphasic	Monophasic
9 in	Biphasic	Biphasic
9 in	Triphasic	Triphasic

Table 1 depicts the results from this experiment. The model made a single error at 9 inches. It predicted a Monophasic sound file as Biphasic. This is likely due to the distance being fairly far and the model taking into account some unnecessary features from the background track. Overall, the system performed proficiently and received an accuracy of 91.66% proving a certain level of resilience towards distance between the device and doppler sound.

Volume Experiment. In the volume experiment, the distance is kept constant at 3 inches with background noise as the volume of the doppler audio playback is changing. The goal of such an experiment is to observe results as the environment around is noisy and the doppler sound output becomes muffled.

Similar to the first experiment, the model made a single error in prediction at 50% volume (42.5 dB). This was likely due to the fact that the volume was

Table 2. Volume experiment results

Volume Level (%)	Prediction	Ground-truth
25%	Monophasic	Monophasic
25%	Biphasic	Biphasic
25%	Triphasic	Triphasic
50%	Monophasic	Monophasic
50%	Triphasic	Biphasic
50%	Triphasic	Triphasic
75%	Monophasic	Monophasic
75%	Biphasic	Biphasic
75%	Triphasic	Triphasic
100%	Monophasic	Monophasic
100%	Biphasic	Biphasic
100%	Triphasic	Triphasic

low and the spectrogram did not gain enough visual features. Other then this, the model received the same accuracy as the distance experiment at 91.66% proving a level of resilience to sound volume varying within a clinical setting. Both of these tests were performed in a preliminary nature and more validation is required within a clinical setting with more testing samples.

4 Conclusion

We have successfully paired a hand-held continuous wave doppler with an all-in-one phasicity predicting hardware component leveraging deep learning. Our baseline validation accuracy was 96.23% and our manually simulated clinical validation accuracy was 91.66%. We have developed a low-cost system which can be used to more objectively assess limb perfusion and enable timely diagnosis of disease at the point-of-care. Significant validation work is required with other doppler machines and in other clinical settings to account for potential technical and dataset bias along with the risk of overfitting our training population. Such validation work could be used to seek regulatory approval to meet the demand for improved point-of-care PAD disease diagnosis and accurate evaluation.

References

1. Nvidia tesla p100: The most advanced data center accelerator. https://www.nvidia.com/en-us/data-center/tesla-p100/
2. Peripheral arterial disease (pad) (September 2020). https://www.cdc.gov/heartdisease/PAD.htm
3. Abadi, M., et al.: Tensorflow: a system for large-scale machine learning. In: 12th {USENIX} Symposium on Operating systems Design and Implementation ({OSDI} 16), pp. 265–283 (2016)

4. Aboyans, V., et al.: Measurement and interpretation of the ankle-brachial index: a scientific statement from the American heart association. Circulation **126**(24), 2890–2909 (2012)
5. Bisong, E.: Google colaboratory. In: Building Machine Learning and Deep Learning Models on Google Cloud Platform, pp. 59–64. Apress, Berkeley, CA (2019). https://doi.org/10.1007/978-1-4842-4470-8_7
6. Brach, J.S., et al.: Incident physical disability in people with lower extremity peripheral arterial disease: the role of cardiovascular disease. J. Am. Geriatr. Soc. **56**(6), 1037–1044 (2008)
7. Criqui, M.H.: Peripheral arterial disease-epidemiological aspects. Vasc. Med. **6**(1_suppl), 3–7 (2001)
8. Criqui, M.H., et al.: Mortality over a period of 10 years in patients with peripheral arterial disease. N. Engl. J. Med. **326**(6), 381–386 (1992)
9. Hirsch, A.T., et al.: Peripheral arterial disease detection, awareness, and treatment in primary care. JAMA **286**(11), 1317–1324 (2001)
10. Hunter, J.D.: Matplotlib: a 2d graphics environment. IEEE Ann. Hist. Comput. **9**(03), 90–95 (2007)
11. Krizhevsky, A., Sutskever, I., Hinton, G.E.: Imagenet classification with deep convolutional neural networks. Adv. Neural. Inf. Process. Syst. **25**, 1097–1105 (2012)
12. LeCun, Y., Bengio, Y., Hinton, G.: Deep learning. Nature **521**(7553), 436–444 (2015)
13. Lee, J.G., et al.: Deep learning in medical imaging: general overview. Korean J. Radiol. **18**(4), 570 (2017)
14. McFee, B., et al.: Librosa: audio and music signal analysis in python. In: Proceedings of the 14th Python in Science Conference, vol. 8, pp. 18–25. Citeseer (2015)
15. Ouriel, K.: Peripheral arterial disease. Lancet **358**(9289), 1257–1264 (2001)
16. Pi, R.: Raspberry pi 3 model b (2015) . https://www.raspberrypi.org
17. Rao, A.: Waveform phasicity prediction from arterial sounds through spectrogram analysis using convolutional neural networks for limb perfusion assessment. arXiv preprint arXiv:2104.09748 (2021)
18. Scissons, R.: Characterizing triphasic, biphasic, and monophasic doppler waveforms: should a simple task be so difficult? J. Diagn. Med. Sonogr. **24**(5), 269–276 (2008)
19. Selesnick, I.W., Burrus, C.S.: Generalized digital butterworth filter design. IEEE Trans. Signal Process. **46**(6), 1688–1694 (1998)
20. Selvaraju, R.R., Cogswell, M., Das, A., Vedantam, R., Parikh, D., Batra, D.: Grad-cam: visual explanations from deep networks via gradient-based localization. In: Proceedings of the IEEE International Conference on Computer Vision, pp. 618–626 (2017)
21. Szegedy, C., Vanhoucke, V., Ioffe, S., Shlens, J., Wojna, Z.: Rethinking the inception architecture for computer vision. In: Proceedings of the IEEE Conference on Computer Vision and Pattern Recognition, pp. 2818–2826 (2016)
22. Tehan, P.E., Chuter, V.H.: Use of hand-held doppler ultrasound examination by podiatrists: a reliability study. J. Foot Ankle Res. **8**(1), 1–7 (2015)
23. Torrey, L., Shavlik, J.: Transfer learning. In: Handbook of Research on Machine Learning Applications and Trends: Algorithms, Methods, and Techniques, pp. 242–264. IGI global (2010)
24. Virtanen, P., et al.: Scipy 1.0: fundamental algorithms for scientific computing in python. Nat. Methods **17**(3), 261–272 (2020)

Learning from Mistakes: An Error-Driven Mechanism to Improve Segmentation Performance Based on Expert Feedback

Siri Willems[1,2](✉), Heleen Bollen[3,4], Julie van der Veen[6], Edmond Sterpin[3,4,5], Wouter Crijns[3,4], Sandra Nuyts[3,4], and Frederik Maes[1,2]

[1] Department of ESAT, Processing Speech and Images (PSI), KU Leuven, 3000 Leuven, Belgium

[2] Medical Imaging Research Center, UZ Leuven, 3000 Leuven, Belgium
{siri.willems,frederik.maes}@kuleuven.be

[3] Department of Oncology, Laboratory of Experimental Radiotherapy, KU Leuven, 3000 Leuven, Belgium

[4] Radiation Oncology, UZ Leuven, 3000 Leuven, Belgium

[5] Department of Radiation Oncology, Cliniques uniersitaires Saint-Luc, UC Louvain, 1200 Woluwe-Saint-Lambert, Belgium

[6] Department of Radiotherapy and Oncology, AZ Sint Maarten, Mechelen, Belgium

Abstract. The goal of this study is to exploit feedback provided by experts that daily interact with a deep learning based segmentation tool to efficiently improve its segmentation performance. A convolutional neural network (CNN) for segmenting organs at risk for head & neck cancer was implemented in clinical practice and its predicted contours were verified, corrected and approved by the radiation oncologists before being used for treatment planning. A second CNN was subsequently trained to predict how the original contours as created by the first CNN should be adapted according to the experts. Three artificial datasets were created from the clinical dataset by introducing different amounts of systematic errors in the original predictions, in addition to the clinically corrected errors. Dice score for brainstem improved from 88% to 90.5% on average for the dataset that was the least adapted, and from 66% to 89.3% on average for the dataset in which the most systematic errors were introduced. For the clinical dataset, final segmentation improved for glottic area and supraglottic larynx compared to the initial predictions. In general, the systematic errors introduced in the contours are easier to learn compared to the clinical corrections by the expert, which are more subtle and subject to observer variability in our clinical dataset.

Keywords: Segmentation · Feedback · Error-driven

1 Introduction

The main goal of radiotherapy is to treat cancer patients by using ionising radiation which damages malignant cells. Hereby it is crucial to avoid healthy tissue,

C. Oyarzun Laura et al. (Eds.): CLIP/DCL/LL-COVID/PPML 2021, LNCS 12969, pp. 68–77, 2021.
https://doi.org/10.1007/978-3-030-90874-4_7

i.e. organs at risk (OARs). Hence precise delineation of tumor tissue and OARs is crucial for treatment planning purposes. These delineation tasks are mainly performed manually leading to long treatment planning times and inter- and intra-observer variability [2,10,14]. Automated segmentation approaches have been widely published in literature. First atlas-based methods were developed for segmentation of both OARs [5] and target volumes [4]. These were followed by deep learning approaches using several variants of convolutional neural networks (CNN) [6,7,12]. While most literature has been focused on OAR segmentation for tumor indications such as head and neck cancer (HNC), lung, and prostate, automated target delineation has been considered as well [3,8]. Automated segmentation proves to be more time efficient and reduces the variability between radiation oncologists [18,19,21].

With the increasing adoption of deep learning methods in clinical practice, it is important that these methods are kept up to date. Due to technological advances in scanner quality and dose delivery methods [13], delineation guidelines and protocols may change or adapt over time. There is a need for a mechanism to update the behavior of the underlying networks by exploiting the clinical feedback that is daily generated when radiation oncologists interact with these learning-based delineation methods.

Different concepts such as transfer learning or reinforcement learning could be relevant to incorporate interactive clinical feedback in a deep learning approach. The goal of transfer learning is to reuse knowledge from one task to solve a second, slightly different task, which can reduce training time and improve accuracy. Specifically for CNNs 2 types of transfer learning can be relevant: instance-based transfer and network-based transfer. For the first type, data samples from the source task could be reused while developing and training a network for the target task. Network-based transfer on the other hand is the most known concept where a network is first pre-trained on the source task, and subsequently trained using part of the pre-trained neural network were knowledge is encoded in the feature representation to solve the target task [15]. Reinforcement learning is rather a trial and error learning technique where the network interacts directly with a certain environment and self teaches over time by making mistakes along the way to reach a predefined goal [11]. The most known reinforcement deep learning applications are computer games [9,17]. Updating the behavior of a segmentation network based on expert feedback can be seen as a reinforcement learning problem. The neural network is the agent that performs a segmentation task and hereby interacts with the clinical environment. The radiation oncologist provides feedback by locally correcting the predicted segmentation if they do not agree with the initial prediction.

We developed such a reinforcement learning concept to efficiently exploit clinical feedback provided as expert-corrected contours in a segmentation network. The feedback is incorporated in a second CNN that is trained using an error-guided mechanism to improve the predictions of the initial CNN by correcting systematic errors that are automatically learned from the expert feedback.

2 Data

The initial dataset (i.e. dataset 1) contained 75 CT scans of HNC patients. One expert radiation oncologist (RO) manually delineated sixteen OARs on each CT scan according to international consensus guidelines of Brouwer et al. [1]. These OARs were brainstem, cochleas, upper esophagus, glottic area, oral cavity, mandible, supraglottic larynx, pharyngeal constrictor muscles, parotid glands, submandibular glands and spinal cord. This dataset was only used to create a 3D convolutional neural network (CNN) for OAR segmentation and was validated and implemented in clinical practice using a client-server system [19,20]. This CNN has been routinely used in our institute to create OAR segmentations for HNC patients, which we refer to as the original predicted contours. These segmentations were carefully reviewed and corrected by several treating ROs within the usual clinical workflow and their revisions verified and approved by a senior RO before being used for radiotherapy treatment planning. This process resulted in a second dataset of 188 new CT scans of recently treated HNC patients for which the original predicted contours and the expert corrected contours are available (i.e. dataset 2). This dataset contains information on the clinically significant errors made by the initially trained CNN and was used to learn to predict these errors. The dataset was randomly divided in a training (124), validation (31) and test set (33). All images were normalised to the interval [0,1] and resampled to have similar voxel sizes of $1 \times 1 \times 3$ mm.

3 Method

Original binary segmentation maps are initially predicted by the clinically implemented 3D CNN. We train a second CNN using to predict whether the initial label of each voxel was correctly predicted or not by the first CNN based on previous expert feedback. The output of this CNN is a binary map with 0 indicating that the voxel label was correctly predicted, and 1 indicating that the voxel label should change value from 0 to 1 or vice versa, see Fig. 1. The final segmentation map $\hat{P}_{fs}$ is then calculated using the function F:

$$F : \hat{P}_{fs} = P_{op} * (1 - \hat{P}_c) + (1 - P_{op}) * \hat{P}_c \tag{1}$$

with P_{op} the original segmentation and $\hat{P}_c$ the predicted corrections. The loss function is calculated over both the predicted correction map and the final segmentation map using the soft Dice similarity coefficient (DSC):

$$L_c = 1 - \frac{2 * \sum(\hat{P}_c * P_c)}{\sum \hat{P}_c + \sum P_c} \tag{2}$$

$$L_{fs} = 1 - \frac{2 * \sum(\hat{P}_{fs} * P_{fs})}{\sum \hat{P}_{fs} + \sum P_{fs}} \tag{3}$$

$$L_{total} = \alpha L_c + (1 - \alpha) L_{fs} \tag{4}$$

with α weighting the relative importance of both terms.

The CNN for error prediction is implemented using a 3D U-net architecture [16], see Fig. 1. The input consists of the CT images and original segmentations cropped around the object of interest. The U-net contains 5 pathways, each consisting of 2 convolutional layers with kernel size (3,3,1) and (3,3,3). The convolutional layers are each followed by batch normalization and a ReLu activation function. Residual connections are added to the convolutional layers. The sigmoid function is used as final activation layer. All networks were implemented with Keras (version 2.2.4) and trained using an NVIDIA GeForce RTX 2080Ti GPU.

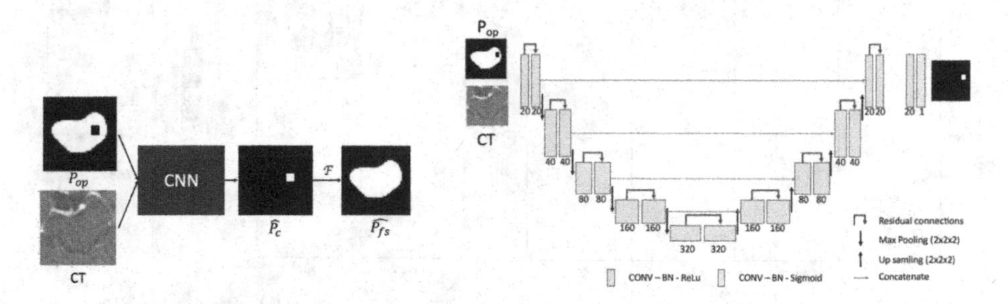

Fig. 1. (Left) Visualisation of the error-driven mechanism with P_{op} being the original prediction, $\hat{P}_c$ the predicted correction to apply and $\hat{P}_f s$ the final segmentation obtained by function F. (Right) U-Net architecture.

4 Experiments and Results

4.1 Proof of Concept: Recovering Systematic Errors

The brainstem was arbitrarily chosen for proof of concept that systematic segmentation errors can be discovered and corrected with our approach. To this end, the original predicted brainstem segmentation was artificially altered by randomly introducing additional, systematic differences with respect to the expert corrected segmentation. These alterations consisted of either removing or adding square patches (creating a hole or a bulb respectively) and either removing or adding (by extrapolation) cranial or caudal slices. Three different brainstem datasets were created in this way according to the randomly chosen parameters in Table 1, with a small (grade 1), an intermediate (grade 2) and a large (grade 3) amount of additional systematic errors introduced between the original prediction (grade 0) and the expert corrected contours (ground truth). The systematic errors added to the dataset accounted for 0%, 31.6%, 45.5% and 76.6% of the overall corrections present in the dataset for grade 0, 1, 2 and 3 respectively. Each dataset can thus be seen as an initial segmentation obtained by different CNNs

Table 1. Different amounts of systematic errors (grade 1–3) were introduced in the brainstem dataset by random alterations made to the original segmentations.

Alteration	Grade 1	Grade 2	Grade 3
Amount of square patches added/removed	0–2	2–4	4–6
Size of patches (mm)	1–3	3	3 or 5
Amount of caudal slices added/removed	0–2	2–4	4–6
Amount of cranial slices added/removed	0–2	2–4	4–6

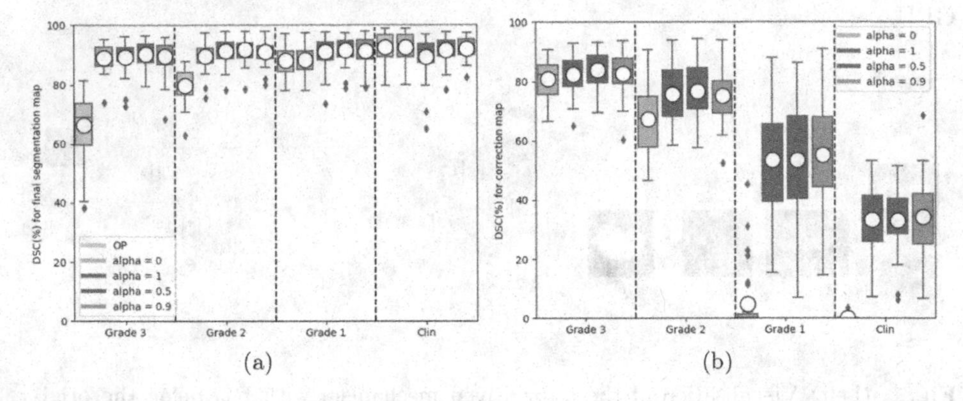

Fig. 2. Boxplots showing DSC (%) computed for (a) the final segmentation and (b) the predicted correction map for each brainstem dataset (grade 0–3) for different loss functions used for training (OP = original prediction).

with different amounts of systematic errors compared to the expert approved final segmentation.

For each dataset, a separate 3D U-net was trained to learn appropriate corrections for that specific dataset (grade 0–3), using weights $\alpha = 1, 0, 0.5, 0.9$ in the loss function (Eq. 4). Error correction performance was evaluated for the test set using DSC computed on the correction map ($\hat{P}_c$) and on the final segmentation map ($\hat{P}_{fs}$), which was compared with the DSC of the original prediction provided as input to the CNN and the expert corrected ground truth. For all datasets except the unaltered clinical dataset (grade 0), the final segmentation improved significantly compared to the original prediction for all evaluated values α, see Fig. 2(a). DSC increased from 66%, 80% and 88% for the original prediction for dataset grade 3, 2 and 1 respectively to over 90% for the final segmentation for each dataset. Figure 2(b) shows the DSC of the correction map itself. Higher DSC values are observed for the datasets with more errors introduced as the required corrections are larger in volume. Weighing both terms L_c and L_{fs} equally in the loss function ($\alpha = 1/2$) gives the best performance on datasets containing larger systematic errors (grade 3), while more subtle errors (grade 0) are somewhat better recovered when giving more weight to the L_c loss.

The CNNs trained with $\alpha = 1/2$ for each dataset (grade 1–3) were subsequently used to predict the corrections for the unaltered segmentations (grade 0) to evaluate whether they are capable of also recovering the more subtle, less systematic clinical corrections. The DSC computed for the correction map is shown in Fig. 3(d). CNNs trained on segmentations with a higher amount of systematic alterations are less able to detect the more subtle corrections applied in clinical practice. Qualitative examples are shown in Fig. 3(a,b,c) for grade 1–3 respectively. These examples illustrate that the proposed approach succeeds at detecting and correcting systematic errors such as holes and missing slices, but that more subtle differences related to corrections made by the expert are more difficult to predict.

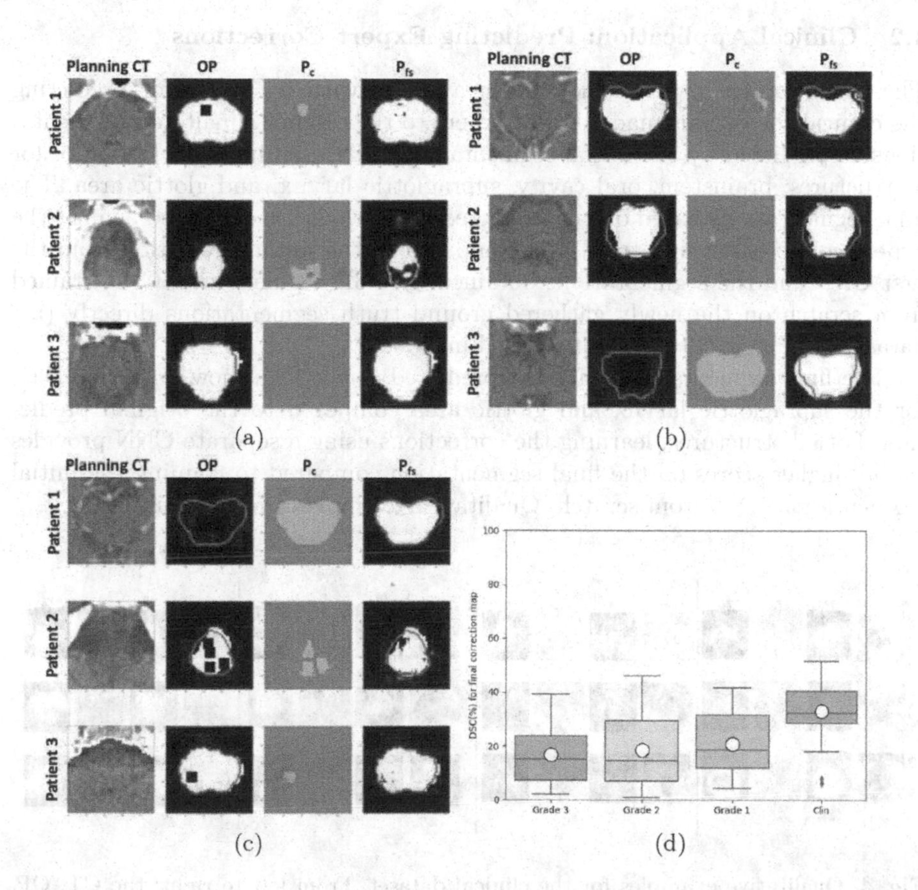

Fig. 3. (a, b, c) Qualitative examples of the corrections predicted for the brainstem datasets with systematic errors of grade 1, 2 and 3 respectively. From left to right: the CT, original prediction, correction map P_c and final segmentation P_{fs}. The ground truth is indicated in red in column 2 and 4. (d) DSC (%) of the correction map predicted by the CNNs trained on each dataset of grade 0–3 with α=1/2 when applied to the original unaltered dataset (grade 0).

Table 2. Mean surface distance (MSD, mm), Hausdorff 95 (HD95, mm) and DSC (%) for the original prediction, the predicted corrections and segmentation trained from scratch against the expert corrected ground truth. Significant differences between original and corrected segmentations are marked in bold.

	Original prediction			Corrected prediction			Trained from scratch		
	MSD	HD95	DSC	MSD	HD95	DSC	MSD	HD95	DSC
Supra. Larynx	1.3(0.9)	4.3(2.3)	79.5(13.8)	1.2(0.6)	**3.6(1.4)**	**83.3(7.5)**	1.7(0.3)	4.1(2.0)	79.7(4.2)
Brainstem	**1.1(0.5)**	3.1(1.8)	**92.5(4.9)**	1.2(0.4)	3.1(1.2)	91.5(4.3)	1.7(0.4)	3.2(1.2)	88.1(4.7)
Oral Cavity	1.5(1.0)	4.1(3.6)	93.5(6.0)	1.7(0.8)	4.1(2.7)	92.4(5.1)	2.3(0.8)	5.0(4.0)	89.7(5.9)
Glottic Area	0.8(0.8)	3.2(3.8)	84.8(12.9)	0.8(0.7)	2.8(3.2)	85.1(11.9)	1.7(2.0)	3.9(4.2)	72.9(17.5)

4.2 Clinical Application: Predicting Expert Corrections

The same approach was used to train a CNN (with $\alpha = 1/2$) for predicting the clinical corrections made by the expert to the original, unaltered segmentations as predicted by the initial segmentation CNN used in clinical practice for 4 structures: brainstem, oral cavity, supraglottic larynx, and glottic area. The final segmentations based on the predicted corrections are evaluated against the expert corrected ground truth and compared to the initial prediction from the first CNN and the segmentations obtained by a 3D segmentation CNN trained from scratch on the newly gathered ground truth segmentations directly (i.e. dataset 2). The results are summarised in Table 2.

The final segmentations with the predicted corrections show improved DSC for the supraglottic larynx and glottic area compared to the original prediction. For all structures, learning the corrections using a separate CNN provides overall higher scores on the final segmentation compared to training the initial segmentation CNN from scratch. Qualitative examples are shown in Fig. 4.

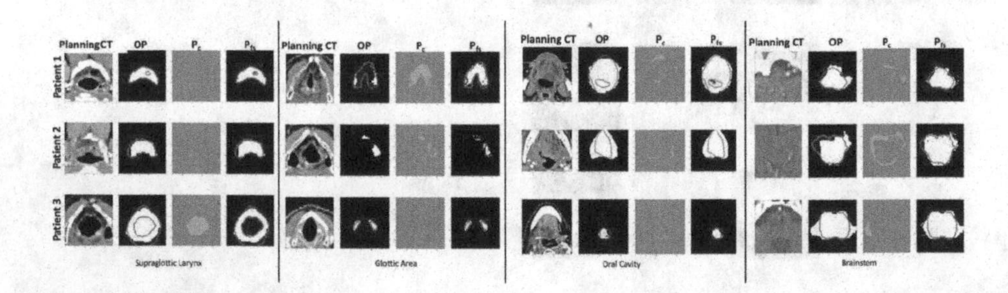

Fig. 4. Qualitative examples for the clinical dataset. From left to right: the CT, OP, P_c and P_{fs}. The ground truth is indicated in red in column 2 and 4. (Color figure online)

5 Discussion and Conclusion

We showed the feasibility of training a CNN with an error-driven mechanism to correct clinically relevant errors in original segmentations generated by a previously trained CNN, by exploiting specific expert feedback collected interactively in clinical practice. We found that systematic errors in the form of artificially introduced holes and missing slices are relatively easy to learn. Similar segmentation performance was achieved irrespective of the amount of such alterations introduced, i.e. corrections for such systematic errors could be learned independently of the quality of the initial predictions. The corrections applied by the clinical expert in routine practice are more subtle, especially when the performance of the originally predicted segmentation is already at a level that makes it clinically useful.

When revising the predicted segmentations, we expect that the expert would systematically correct the systematic errors in the predictions made by the initial CNN, for which there is evidence in the image for detecting and correcting these errors. We hypothesize that these systematic patterns could in principle be learned by a second CNN that has the original prediction and the image itself as input, as we demonstrated for the artificially created datasets in the first experiment. However, corrections of the initially predicted contours for the images in our clinical dataset were made by multiple radiation oncologists and although all were reviewed and approved by the same senior radiation oncologist, some observer variability and randomness in the corrections applied by the experts to the automatically generated initial contours is to be expected due to ambiguity in the images, although less than for pure manual delineation. We previously found the similarity between 2 observers correcting independently the same predictions on the same clinical images to be around 70%, 96%, 93% and 90% for brainstem, oral cavity, supraglottic larynx and glottic area respectively [19]. The feedback obtained from the experts is thus not perfect but in part unreliable and insignificant. The CNN that we train to predict the required corrections to improve the performance of the initial segmentation CNN, needs to be able to discard such irrelevant feedback and to focus on systematic error patterns that appear correlated to the image texture or the shape of the individual segmentations.

The corrected segmentations of the clinical dataset reached a performance which is comparable to other studies regarding OAR segmentation for HNC patients [7,12]. The largest increase in performance was seen for structures that were initially less well segmented by the original CNN, in particular the supraglottic larynx. The glottic area improved slightly but not significantly compared to its initial segmentation. The oral cavity and brainstem were already fairly well segmented with the original CNN, making their corrections even more subtle and more difficult to learn. Note that the initial CNN is corrected and thus automatically reaches a high conformity which results in a slight bias for the Dice score in favor of the original CNN. This could be improved by gathering manual segmentations of the test set in the future and compare each method with these manual ground truth segmentations. In conclusion, we presented a new learning concept to improve the performance of a segmentation CNN by efficiently exploiting feedback provided in clinical practice by experts that directly interact with the CNN. The concept showed promising results on artificial datasets containing systematic errors. The advantage of this approach is that the original CNN is treated as a black box and need not be retrained as more data become available or segmentation protocols are updated. In future work, we aim to explore how the same error-driven mechanism can be used to provide clues to the expert about the reliability of the provided automated predictions.

Acknowledgments. Siri Willems and Heleen Bollen are supported by a Ph.D. fellowship of the research foundation – Flanders (FWO).

References

1. Brouwer, C.L., et al.: Ct-based delineation of organs at risk in the head and neck region: DAHANCA, EORTC, GORTEC, HKNPCSG, NCIC CTG, NCRI, NRG oncology and TROG consensus guidelines. Radiother. Oncol. **117**(1), 83–90 (2015)
2. Brouwer, C.L., et al.: 3D variation in delineation of head and neck organs at risk. Radiat. Oncol. **7**(1), 1–10 (2012)
3. Cardenas, C.E., et al.: Deep learning algorithm for auto-delineation of high-risk oropharyngeal clinical target volumes with built-in dice similarity coefficient parameter optimization function. Int. J. Radiat. Oncol. *Biol.* Phys. **101**(2), 468–478 (2018)
4. Commowick, O., Grégoire, V., Malandain, G.: Atlas-based delineation of lymph node levels in head and neck computed tomography images. Radiother. Oncol. **87**(2), 281–289 (2008)
5. Daisne, J.F., Blumhofer, A.: Atlas-based automatic segmentation of head and neck organs at risk and nodal target volumes: a clinical validation. Radiat. Oncol. **8**(1), 1–11 (2013)
6. van Dijk, L.V., et al.: Improving automatic delineation for head and neck organs at risk by deep learning contouring. Radiother. Oncol. **142**, 115–123 (2020)
7. Ibragimov, B., Xing, L.: Segmentation of organs-at-risks in head and neck CT images using convolutional neural networks. Med. Phys. **44**(2), 547–557 (2017)
8. Men, K., et al.: Fully automatic and robust segmentation of the clinical target volume for radiotherapy of breast cancer using big data and deep learning. Physica Med. **50**, 13–19 (2018)
9. Mnih, V., et al.: Playing atari with deep reinforcement learning. arXiv preprint arXiv:1312.5602 (2013)
10. Mukesh, M., et al.: Interobserver variation in clinical target volume and organs at risk segmentation in post-parotidectomy radiotherapy: can segmentation protocols help? Br. J. Radiol. **85**(1016), e530–e536 (2012)
11. Nguyen, T.T., Nguyen, N.D., Nahavandi, S.: Deep reinforcement learning for multiagent systems: a review of challenges, solutions, and applications. IEEE Trans. Cybern. **50**(9), 3826–3839 (2020)
12. Nikolov, S., et al.: Deep learning to achieve clinically applicable segmentation of head and neck anatomy for radiotherapy. arXiv preprint arXiv:1809.04430 (2018)
13. Palma, D., et al.: Volumetric modulated arc therapy for delivery of prostate radiotherapy: comparison with intensity-modulated radiotherapy and three-dimensional conformal radiotherapy. Int. J. Radiat. Oncol. *Biol.* Phys. **72**(4), 996–1001 (2008)

14. Piotrowski, T., et al.: Impact of the intra-and inter-observer variability in the delineation of parotid glands on the dose calculation during head and neck helical tomotherapy. Technol. Cancer Res. Treat. **14**(4), 467–474 (2015)
15. Ribani, R., Marengoni, M.: A survey of transfer learning for convolutional neural networks. In: 2019 32nd SIBGRAPI Conference on Graphics, Patterns and Images Tutorials (SIBGRAPI-T), pp. 47–57. IEEE (2019)
16. Ronneberger, O., Fischer, P., Brox, T.: U-Net: convolutional networks for biomedical image segmentation. In: Navab, N., Hornegger, J., Wells, W.M., Frangi, A.F. (eds.) MICCAI 2015. LNCS, vol. 9351, pp. 234–241. Springer, Cham (2015). https://doi.org/10.1007/978-3-319-24574-4_28
17. Taylor, M.E.: Teaching reinforcement learning with Mario: an argument and case study. In: Proceedings of the 2011 AAAI Symposium Educational Advances in Artificial Intelligence (2011)
18. Van der Veen, J., Willems, S., Bollen, H., Maes, F., Nuyts, S.: Deep learning for elective neck delineation: more consistent and time efficient. Radiother. Oncol. **153**, 180–188 (2020)
19. Van der Veen, J., et al.: Benefits of deep learning for delineation of organs at risk in head and neck cancer. Radiother. Oncol. **138**, 68–74 (2019)
20. Willems, S., et al.: Clinical implementation of DeepVoxNet for auto-delineation of organs at risk in head and neck cancer patients in radiotherapy. In: Stoyanov, D. (ed.) CARE/CLIP/OR 2.0/ISIC -2018. LNCS, vol. 11041, pp. 223–232. Springer, Cham (2018). https://doi.org/10.1007/978-3-030-01201-4_24
21. Wong, J., et al.: Comparing deep learning-based auto-segmentation of organs at risk and clinical target volumes to expert inter-observer variability in radiotherapy planning. Radiother. Oncol. **144**, 152–158 (2020)

TMJOAI: An Artificial Web-Based Intelligence Tool for Early Diagnosis of the Temporomandibular Joint Osteoarthritis

Celia Le[1(✉)], Romain Deleat-Besson[1], Najla Al Turkestani[1], Lucia Cevidanes[1], Jonas Bianchi[3], Winston Zhang[1], Marcela Gurgel[1], Hina Shah[1], Juan Prieto[2], and Tengfei Li[2]

[1] University of Michigan, Ann Arbor, MI 48109, USA
celiale@umich.edu
[2] University of North Carolina, Chapel Hill, NC, USA
[3] University of the Pacific, San Francisco, CA, USA

Abstract. Osteoarthritis is a chronic disease that affects the temporomandibular joint (TMJ), causing chronic pain and disability. To diagnose patients suffering from this disease before advanced degradation of the bone, we developed a diagnostic tool called TMJOAI. This machine learning based algorithm is capable of classifying the health status TMJ in of patients using 52 clinical, biological and jaw condyle radiomic markers. The TMJOAI includes three parts. the feature preparation, selection and model evaluation. Feature generation includes the choice of radiomic features (condylar trabecular bone or mandibular fossa), the histogram matching of the images prior to the extraction of the radiomic markers, the generation of feature pairwise interaction, etc.; the feature selection are based on the p-values or AUCs of single features using the training data; the model evaluation compares multiple machine learning algorithms (e.g. regression-based, tree-based and boosting algorithms) from 10 times 5-fold cross validation. The best performance was achieved with averaging the predictions of XGBoost and LightGBM models; and the inclusion of 32 additional markers from the mandibular fossa of the joint improved the AUC prediction performance from 0.83 to 0.88. After cross-validation and testing, the tools presented here have been deployed on an open-source, web-based system, making it accessible to clinicians. TMJOAI allows users to add data and automatically train and update the machine learning models, and therefore improve their performance.

Keywords: Machine learning · Early diagnosis · Osteoarthritis

1 Introduction

Temporomandibular joints (TMJ) are small joints that connect the lower jaw (mandible) to the skull. They are susceptible to suffer from disorders causing

Supported by NIDCR DE024550 and AAOF Dewel Biomedical research Award.

C. Oyarzun Laura et al. (Eds.): CLIP/DCL/LL-COVID/PPML 2021, LNCS 12969, pp. 78–87, 2021.
https://doi.org/10.1007/978-3-030-90874-4_8

recurrent or chronic pain and dysfunction, making them among the most common causes of facial pain [20]. Osteoarthritis (OA) is the most common form of arthritis which is a condition affecting over 50 million US adults and leading to chronic disability and alteration in the structure of the joints [5]. An early diagnosis could help in reducing the destruction of the bone by slowing the disease's progression, which is essential for suffering patients since there is to this day no cure for OA, other than reducing its symptoms [19,23].

Integration of clinical, biological and radiomic markers have been shown in previous studies to contribute to a precise diagnosis of TMJ osteoarthritis (TMJOA) [2,6]. Biological markers are obtained by quantifying protein levels in synovial fluid, serum, and saliva, while imaging markers are obtained by acquiring cone-beam computed tomography (CBCT) images of the TMJ region [10,22].

To meet the need for a robust early diagnosis tool, we developed TMJOAI, an artificial intelligence-based tool, using machine learning algorithms such as regression trees, which have been shown to help finding correlation between variables in order to predict OA [2,12,13,17], or other diseases [9]. We compared different algorithms to find the most efficient one in classifying patients' health status.

The dataset used for training those machine learning models is detailed in Sect. 2. The method and the different algorithms are described in Sect. 3. The results of the performance experiments are presented in Sect. 4, and the conclusions are drawn in Sect. 5.

2 Dataset

Our dataset consisted in 92 subjects, 46 suffering from TMJ OA, and 46 healthy controls, making it a balanced dataset. Moreover, the OA and the control groups were age and sex matched. The data acquisition protocol was the same for all subjects. We obtained the values of 52 markers: 2 demographic values, 13 protein level values from serum, 12 protein level values from saliva, 5 clinical features evaluating the pain and 20 imaging features representing the grey-level values of the region of interest. The imaging features originate from the lateral region of the trabecular bone of the condyle, which is the lower part of the TMJ. For more robust radiomic markers, we also tested 32 mandibular fossa radiomic features, which forms the upper part of the TMJ. Therefore, our training dataset had a total of 3828 features and interactions features.

Those values were merged and stored in a csv file, which is given as input for our machine learning algorithms.

3 Proposed Methods

3.1 Feature Selection

Out of the 52 features composing our dataset, we calculated the interaction value between each of them by multiplying them, resulting in 1326 additional features, for a total of 1378 features. We then computed the Area Under the Receiver Operating Curve (AUC) of each feature to evaluate their relevance and select the features with higher AUC (Fig. 1), which have better performance in the classification task of the TMJ health status.

In addition, we calculated the correlation between each pair of the selected features. For those highly auto-correlated pairs we chose the one with the highest AUC and excluded the others.

This correlation-based selection helps to reduce the complexity of the model and prevent overfitting.

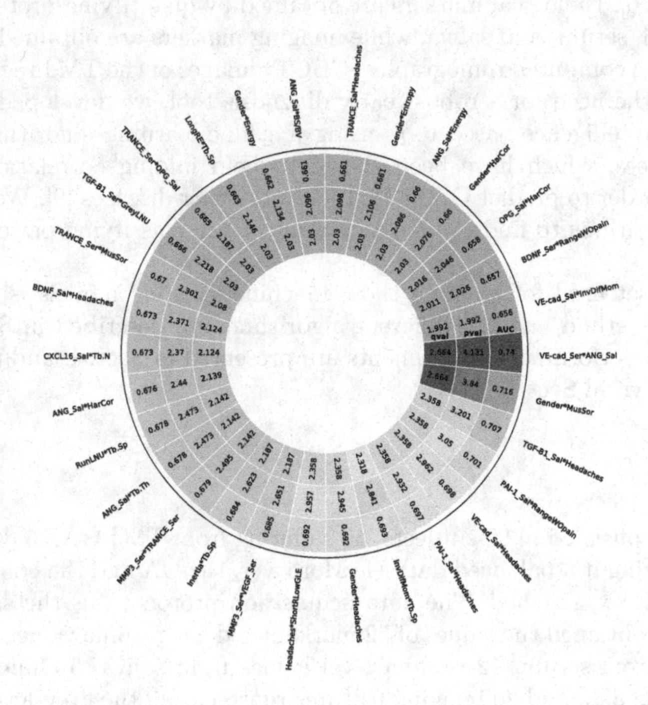

Fig. 1. Circular plot of the AUC, $-log(pvalue)$ and $-log(qvalue)$ of the features (outside to inside)

3.2 Comparison of Multiple Machine Learning Algorithms

We divided our dataset into 5-folds to perform a leave-one-out cross validation, using stratified folds to keep the balance between the OA and control groups in each fold. We repeated the operation 10-times to bypass the sampling bias from the random subdivision of the train-test folds, by using different seeds for randomly creating the folds of the cross validation. The seeds were fixed to allow the reproducibility of the algorithm. This resulted in 50 models for each of the trained algorithms.

We compared 5 different algorithms to select which performed the best: Random Forest, XGBoost, LightGBM, Ridge and Logistic Regressions. Those algorithms are well-indicated to solve both a regression or a classification problem, such as the diagnosis of a disease, like in the present case [15,16]. Regression models weigh the prediction and provide outputs as a probability of disease.

Regression Analysis. Regression analysis designates a collection of methods used in machine learning to predict a dependent variable (the health status of a patient) from multiple independent variables or predictors (our 52 features values and their associated interactions). These methods use different functions during the regression.

The ridge regression uses a linear regression, to which we add a ridge regression penalty to solve the problem:

$$L(x, y) = min(\sum (y_i - w_i x_i)^2 + \lambda \sum w_i^2) \tag{1}$$

where $\lambda = [1, 10]$, and its optimal value is determined by doing a nested 5-folds cross validation.

Whereas the logistic regression uses a sigmoid function:

$$f(x) = \frac{1}{1 + e^{-x}} \tag{2}$$

Regression Trees. Regression trees are an underlying branch of regression analysis. As decision trees, they use a binary recursive partitioning to split the data into 2 branches (diseased or healthy) according to the value of a random subset of the features. The tree will keep growing on the branch that minimizes the mean squared error:

$$1/n * \sum (y - prediction)^2 \tag{3}$$

A Random Forest is composed of an ensemble of regression trees (500 in our case), where each tree is trained using only a subdivision of the features and a subdivision of the dataset. The final prediction is the average of the predictions from all the trees. It has shown its ability to solve both classification and regression problems [3].

Gradient-Boosted Regression Trees. eXtreme Gradient Boosted machine (XGBoost) [7] and Light Gradient Boosted Machine (LightGBM) [14] are gradient-boosted tree algorithms. They both originate from Random Forest but vary in their tree's growth pattern. Moreover, gradient boosting methods combine the predictions while building the trees instead of doing it at the end of the process.

The gradient-boosted methods have the difference that XGBoost has a level-wise tree growth whereas LightGBM has a leaf-wise tree growth and uses Gradient-based One-Side Sampling (GOSS).

They also offer the possibility to tune more parameters, and if done properly demonstrated superior performances than Random Forest [1,8,11,18,21].

For the gradient-boosted algorithms, we performed grid search to determine the optimal value of the different hyperparameters:

- Learning rate: Step size.
- Subsample/Bagging fraction: The fraction of cases to be used in each tree, to prevent overfitting, applied once in every boosting iteration.
- Colsample by tree/Feature fraction: The fraction of features to be used in each tree, to prevent overfitting, applied once every tree.
- Minimum child weight: The minimum weight of a branch, under which the branch will stop growing to prevent the overfitting of the tree.
- Max depth: Maximum depth to which the trees will grow, a low value is set to prevent overfitting.
- Number of estimators: Number of trees in the forest, combined with a nested 5-folds cross validation to find the optimal number of trees.

The range of values used during the grid search and the values selected to train the models are reported in Table 1.

Table 1. Range and values of hyperparameters.

	Learning rate	Subsample	Colsample	Min weight	Max depth	Estimators
Range	[0.01, 0.001]	0.5, 0.7	0.5, 0.7	1, 2	[1, 10]	[1000, 10000]
XGBoost	0.01	0.5	0.7	2	1	5000
LightGBM	0.005	0.5	0.7	2	1	1000

3.3 Histogram Matching

The differences among the grey-levels of the trabecular imaging data were adjusted by histogram matching to control subjects. All images were each matched to 5 different references from the control subjects. The mean value of each radiomic feature was then used as a training data.

4 Experimental Results

4.1 Experiments

We obtained a probability of TMJ OA for each patient from our 5-folds cross validation. We applied a threshold of 0.5 to determine the final health status prediction of the model and calculated the following metrics to evaluate the performances of the model: accuracy, precision and recall for OA group (1) and control group (0), F1-score and AUC.

Table 2. Comparison of metrics for the different algorithms.

Models	Accuracy	Precision1	Precision0	Recall1	Recall0	F1-Score	AUC
RandomForest	0.705	0.710	0.704	0.696	0.715	0.701	0.763
Std	± 0.038	± 0.039	± 0.041	± 0.069	± 0.053	± 0.048	± 0.034
XGBoost	0.714	0.717	**0.712**	**0.709**	0.720	**0.712**	**0.780**
Std	± 0.029	± 0.033	± 0.030	± 0.037	± 0.039	± 0.030	± 0.038
LightGBM	**0.738**	**0.740**	**0.738**	**0.735**	**0.741**	**0.737**	**0.802**
Std	± 0.038	± 0.037	± 0.043	± 0.053	± 0.040	± 0.041	± 0.040
Ridge	0.670	0.678	0.663	0.646	0.693	0.661	0.704
Std	± 0.029	± 0.029	± 0.032	± 0.057	± 0.040	± 0.038	± 0.033
Logistic	**0.715**	**0.740**	0.695	0.663	**0.767**	0.699	0.775
Std	± 0.027	± 0.027	± 0.028	± 0.039	± 0.025	± 0.031	± 0.022

4.2 Algorithm Comparison Results

We averaged the metrics of the 50 models for each algorithm and reported the results in Table 2.

We have concluded from this experiment that LightGBM and XGBoost are the most efficient diagnostic methods since they had the highest AUCs and F1-scores. Consequently, we decided to combine the models by averaging the 50 XGBoost and the 50 LightGBM predictions to give the final classification.

Due to the small size of the dataset, we decided not to use a testing dataset, and thus, in order to evaluate the performance of the model, the prediction of each subject only used the models out of the 100 models that were trained without this subject. The classification was therefore made by 20 different models (10 XGBoost and 10 LightGBM) for each of the folds, and compared to the true health status.

As we can see in the Table 3, by combining XGBoost and LightGBM models, we obtain a final model more efficient than each of them separately.

Table 3. Comparison of metrics for the final model.

Models	Accuracy	Precision1	Precision0	Recall1	Recall0	F1-Score	AUC
Mean XGB+LGBM	0.726	0.728	0.725	0.722	0.730	0.725	0.791
Models combination	**0.761**	**0.761**	**0.761**	**0.761**	**0.761**	**0.761**	**0.831**

4.3 Histogram Matching and Mandibular Fossa Features Results

Figure 2 shows the AUC of the interaction features before and after histogram matching. In Fig. 2b, the condylar trabecular features (in cyan) provided a high contrast in their interactions with each of the features; some of them demonstrated a significantly higher AUC. By adding the mandibular fossa features, we observed that a substantial number of these features, particularly those combined with the clinic and condyle trabecular features, had an overall higher AUC than the other features.

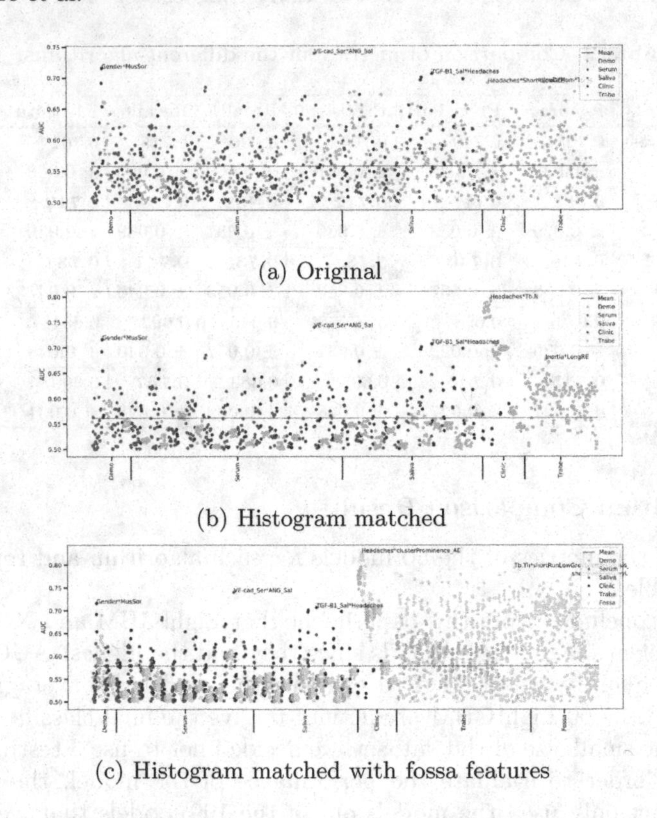

(a) Original

(b) Histogram matched

(c) Histogram matched with fossa features

Fig. 2. AUC of the interaction features.

When we compared the results of the training with and without histogram matched features, including mandibular fossa features in Table 4, using the same training parameters, we noticed that the AUC is slightly higher with histogram matched features, and improved considerably by adding the mandibular fossa features.

Table 4. Comparison of metrics for the final model with histogram matched imaging/trabecular features.

Models	Accuracy	Precision1	Precision0	Recall1	Recall0	F1-Score	AUC
Original	0.761	0.761	0.761	**0.761**	0.761	0.761	0.831
Histogram Matched	0.750	0.767	0.735	0.717	0.783	0.742	0.835
HM+Fossa	**0.794**	**0.814**	**0.776**	**0.761**	**0.826**	**0.787**	**0.882**

The fossa features improved the performances of the final model, in particular the AUC by 0.5 and the F1-score by 0.25. These findings demonstrate the importance of including mandibular fossa features as well as the trabecular ones.

4.4 Deployment

Web-Based System. Our aim was to make this tool available for clinicians without the need for a developer to run it. We uploaded the TMJOAI tool into a docker image, running on a web-based system called Data Storage for Computation and Integration (DSCI) [4].

The code is written in python which makes it easy to deploy, since python environments are available in Docker and are highly maintainable.

Prediction. The prediction algorithm takes the values of the clinical, biological and radiomic markers of one or multiple patients as an input in a single csv file. Then, it returns a csv file with the predicted diagnosis for each patient.

Since having the value of all of the features can be complex, it is possible to maintain the efficiency of the prediction with part of the data. However, absence of essential features can impact the accuracy of the diagnosis (Fig. 3).

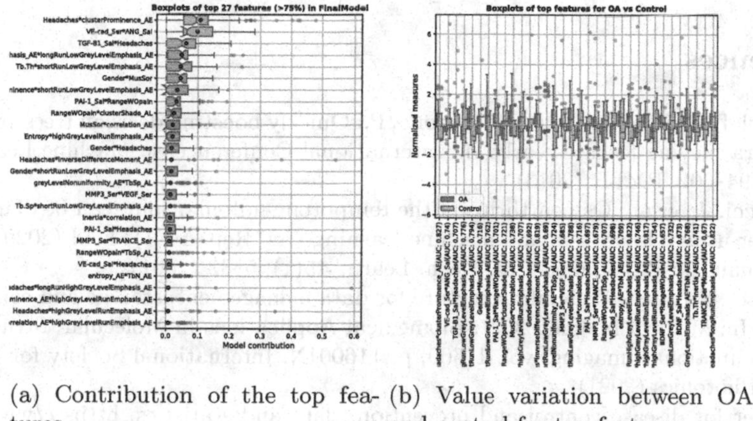

(a) Contribution of the top features (b) Value variation between OA and control for top features

Fig. 3. Statistics of the top features

Training. Since our dataset only included 92 patients, our goal was to enable the addition of new cases to the training dataset. This will improve the TMJOAI tool by making the model more robust. We implemented the functionality to add data and automatically retraining the models and therefore improve their performances. In that way, augmenting the dataset and improving the tool is fast (about 20 min), effortless and doesn't require many of developer supervision.

5 Conclusion

By combining two efficient regression tree algorithms, we have been able to develop a tool capable of predicting TMJ OA before the appearance of symptoms, by using clinical, biological and radiomic markers. This early diagnosis could help in applying earlier intervention methods that prevent progressive destruction of the TMJ bone.

These experiments have shown the efficiency of gradient-boosted methods, as well as associating multiple models, using regression to give more importance to some predictions by giving as output a probability instead of a classification.

However, only 92 cases have been used for this study, therefore this tool can still be improved by adding new patients to our dataset to make the model more robust, hence the necessity of being able to easily add data and train quickly the model.

Finally, in this study, we successfully deployed the artificial web-based intelligence tool TMJOAI towards an early diagnosis of the Temporomandibular Joint Osteoarthritis.

References

1. Appel, R., Fuchs, T., Dollár, P., Perona, P.: Quickly boosting decision trees-pruning underachieving features early. In: International Conference on Machine Learning, pp. 594–602. PMLR (2013)
2. Bianchi, J., et al.: Osteoarthritis of the temporomandibular joint can be diagnosed earlier using biomarkers and machine learning. Sci. Rep. **10**(1), 1–14 (2020)
3. Breiman, L.: Random forests. Mach. Learn. **45**(1), 5–32 (2001)
4. Brosset, S., et al.: Web infrastructure for data management, storage and computation. In: Medical Imaging 2021: Biomedical Applications in Molecular, Structural, and Functional Imaging, vol. 11600, p. 116001N. International Society for Optics and Photonics (2021)
5. Center for disease control and prevention. data and statistics. https://www.cdc.gov/arthritis/data_statistics/index.htm. Accessed July 2021
6. Cevidanes, L.H., et al.: 3D osteoarthritic changes in TMJ condylar morphology correlates with specific systemic and local biomarkers of disease. Osteoarthr. Cartil. **22**(10), 1657–1667 (2014)
7. Chen, T., Guestrin, C.: XGBoost: a scalable tree boosting system. In: Proceedings of the 22nd ACM SIGKDD International Conference on Knowledge Discovery and Data Mining, pp. 785–794. ACM, New York (2016). **10**(2939672.2939785)
8. Chen, T., Li, H., Yang, Q., Yu, Y.: General functional matrix factorization using gradient boosting. In: International Conference on Machine Learning, pp. 436–444. PMLR (2013)
9. Cosma, G., Brown, D., Archer, M., Khan, M., Pockley, A.G.: A survey on computational intelligence approaches for predictive modeling in prostate cancer. Expert Syst. Appl. **70**, 1–19 (2017)
10. Ebrahim, F.H., et al.: Accuracy of biomarkers obtained from cone beam computed tomography in assessing the internal trabecular structure of the mandibular condyle. Oral Surg. Oral Med. Oral Pathol. Oral Radiol. **124**(6), 588–599 (2017)

11. Friedman, J.H.: Greedy function approximation: a gradient boosting machine. Ann. Stat. **29**, 1189–1232 (2001)
12. Heard, B.J., Rosvold, J.M., Fritzler, M.J., El-Gabalawy, H., Wiley, J.P., Krawetz, R.J.: A computational method to differentiate normal individuals, osteoarthritis and rheumatoid arthritis patients using serum biomarkers. J. R. Soc. Interface **11**(97), 20140428 (2014)
13. Jamshidi, A., Pelletier, J.P., Martel-Pelletier, J.: Machine-learning-based patient-specific prediction models for knee osteoarthritis. Nat. Rev. Rheumatol. **15**(1), 49–60 (2019)
14. Ke, G., et al.: LightGBM: a highly efficient gradient boosting decision tree. Adv. Neural Inf. Process. Syst. **30**, 3146–3154 (2017)
15. Kuo, D.E., et al.: Gradient boosted decision tree classification of endophthalmitis versus uveitis and lymphoma from aqueous and vitreous IL-6 and IL-10 levels. J. Ocul. Pharmacol. Ther. **33**(4), 319–324 (2017)
16. Kuo, D.E., et al.: Logistic regression classification of primary vitreoretinal lymphoma versus uveitis by interleukin 6 and interleukin 10 levels. Ophthalmology **127**(7), 956–962 (2020)
17. Lazzarini, N., et al.: A machine learning approach for the identification of new biomarkers for knee osteoarthritis development in overweight and obese women. Osteoarthr. Cartil. **25**(12), 2014–2021 (2017)
18. Li, P., Wu, Q., Burges, C.: McRank: learning to rank using multiple classification and gradient boosting. Adv. Neural Inf. Process. Syst. **20**, 897–904 (2007)
19. Liu, Y., et al.: Multiple treatment meta-analysis of intra-articular injection for temporomandibular osteoarthritis. J. Oral Maxillofac. Surg. **78**(3), 373-e1 (2020)
20. National institute of dental and craniofacial research. facial pain. https://www.nidcr.nih.gov/research/data-statistics/facial-pain. Accessed July 2021
21. Oguz, B.U., Shinohara, R.T., Yushkevich, P.A., Oguz, I.: Gradient boosted trees for corrective learning. In: Wang, Q., Shi, Y., Suk, H.-I., Suzuki, K. (eds.) MLMI 2017. LNCS, vol. 10541, pp. 203–211. Springer, Cham (2017). https://doi.org/10.1007/978-3-319-67389-9_24
22. Paniagua, B., et al.: Validation of CBCT for the computation of textural biomarkers. In: Medical Imaging 2015: Biomedical Applications in Molecular, Structural, and Functional Imaging, vol. 9417, p. 94171B. International Society for Optics and Photonics (2015)
23. Wang, X., Zhang, J., Gan, Y., Zhou, Y.: Current understanding of pathogenesis and treatment of TMJ osteoarthritis. J. Dent. Res. **94**(5), 666–673 (2015)

COVID-19 Infection Segmentation from Chest CT Images Based on Scale Uncertainty

Masahiro Oda[1,2(✉)], Tong Zheng[2], Yuichiro Hayashi[2], Yoshito Otake[3,4], Masahiro Hashimoto[5], Toshiaki Akashi[6], Shigeki Aoki[6], and Kensaku Mori[1,2,4]

[1] Information and Communications, Nagoya University, Furo-cho, Chikusa-ku, Nagoya, Aichi 4648601, Japan
moda@i.nagoya-u.ac.jp

[2] Graduate School of Informatics, Nagoya University, Nagoya, Aichi, Japan

[3] Graduate School of Science and Technology, Nara Institute of Science and Technology, Nara, Japan

[4] Research Center for Medical Bigdata, National Institute of Informatics, Tokyo, Japan

[5] Department of Radiology, Keio University School of Medicine, Tokyo, Japan

[6] Department of Radiology, Juntendo University, Tokyo, Japan

Abstract. This paper proposes a segmentation method of infection regions in the lung from CT volumes of COVID-19 patients. COVID-19 spread worldwide, causing many infected patients and deaths. CT image-based diagnosis of COVID-19 can provide quick and accurate diagnosis results. An automated segmentation method of infection regions in the lung provides a quantitative criterion for diagnosis. Previous methods employ whole 2D image or 3D volume-based processes. Infection regions have a considerable variation in their sizes. Such processes easily miss small infection regions. Patch-based process is effective for segmenting small targets. However, selecting the appropriate patch size is difficult in infection region segmentation. We utilize *the scale uncertainty* among various receptive field sizes of a segmentation FCN to obtain infection regions. The receptive field sizes can be defined as the patch size and the resolution of volumes where patches are clipped from. This paper proposes an infection segmentation network (ISNet) that performs patch-based segmentation and a scale uncertainty-aware prediction aggregation method that refines the segmentation result. We design ISNet to segment infection regions that have various intensity values. ISNet has multiple encoding paths to process patch volumes normalized by multiple intensity ranges. We collect prediction results generated by ISNets having various receptive field sizes. Scale uncertainty among the prediction results is extracted by the prediction aggregation method. We use an aggregation FCN to generate a refined segmentation result considering scale uncertainty among the predictions. In our experiments using 199 chest CT volumes of COVID-19 cases, the prediction aggregation method improved the dice similarity score from 47.6% to 62.1%.

Keywords: COVID-19 · Infection segmentation · Scale uncertainty

C. Oyarzun Laura et al. (Eds.): CLIP/DCL/LL-COVID/PPML 2021, LNCS 12969, pp. 88–97, 2021.
https://doi.org/10.1007/978-3-030-90874-4_9

1 Introduction

Novel coronavirus disease 2019 (COVID-19) spread worldwide, causing many infected patients and deaths. The total number of cases and deaths related to COVID-19 are more than 212 million and 4.4 million in the world [1]. Because of the rapid increase of COVID-19 patients, medical institutions suffer from a human resources shortage. To prevent further infection, a quick inspection method for COVID-19 infection is pressing required. Such quick inspection enables providing appropriate treatments to patients and curbs the spread of COVID-19. Reverse transcriptase-polymerase chain reaction (RT-PCR) testing is used as an inspection method of COVID-19 cases. However, it takes some hours to give a diagnosis result. Furthermore, its sensitivity is not high, ranging from 42% to 71% [2]. As another choice of COVID-19 cases, CT image-based diagnosis is helpful. The sensitivity of CT image-based COVID-19 diagnosis is reported as 97% [3]. Furthermore, a CT scan takes only some minutes. A CT image-based computer-aided diagnosis (CAD) system for COVID-19 is expected to provide a quick and accurate diagnosis to patients. For such CAD systems, a quantitative analysis method of the lung condition is essential. Ground-glass opacities (GGOs) and consolidations are commonly found in the lung of viral pneumonia cases, including COVID-19. We call them *infection regions*. Automatic segmentation of them is an essential component of CAD systems.

Related Work of COVID-19 Segmentation: Previously, deep learning-based automatic segmentation methods of infection regions from CT volumes of COVID-19 cases were proposed [4–8]. Fan et al. [4] proposed an infection region segmentation method using the Inf-Net. The Inf-Net utilizes reverse attention and edge attention to learn features to differentiate infection and other regions. However, because they employ 2D image-based process, 3D positional information is not utilized in their segmentation method. Other papers also employ 2D image-based process [5–7]. Yan et al. [8] proposed a fully convolutional network (FCN) to segment infection and normal regions in the lung. The FCN has contrast enhancement branches to extract features of infection regions that have various intensities. However, because contrast information of segmentation targets is not explicitly provided to the FCN, the contrast enhancement branches' contribution to improving segmentation accuracy is limited.

Scale Uncertainty on Patch-based Process: Infection regions contain many small regions. Segmentation processes hardly segment them from whole 2D slice image or 3D volume as performed in the previous methods [4,8]. To segment such small regions, a patch-based approach is practical. Patch-based approach is commonly employed in segmentation methods from images of large data size such as 3D CT volume [9–11] or pathological images [12–15]. The approach is advantageous to perform deep learning-based segmentation under the limitation of GPU memory size. In patch-based approaches, patch size is an essential factor of segmentation accuracy. The patch size defines the size of the receptive field of segmentation models. Also, original images or volumes can be scaled before patch clipping to change the receptive field size. In summary, (a) *the resolution of original volume (VRes)* and (b) *the size of patch (PSize)* are essential factors for the segmentation accuracy in patch-based approaches. In a multi-organ segmentation method [9], the use of a relatively large PSize resulted in the achievement of high segmentation accuracies among large organs (liver, spleen, and stomach). However, their segmentation accuracy of small organs (artery, vein, and pancreas) was low. Other paper [10] reported that the use of small PSize is effective for small organ (artery) segmentation. VRes and PSize should be selected to patch covers the segmentation target from their results.

In infection region segmentation, selecting appropriate VRes and PSize are difficult because the sizes of infection regions are different for each region. If we apply a segmentation process using multiple VRess and PSizes, we can obtain multiple prediction maps having variation among them. The variation can be considered as *uncertainty among scales.* The scale uncertainty represents useful information to obtain an accurate segmentation result. Scale uncertainty-aware aggregation process of multiple prediction maps is essential for segmenting infection regions with various sizes.

Proposed Method and Contributions: We present an infection region segmentation method from a chest CT volume of a COVID-19 patient. We developed a patch-based FCN for infection region segmentation called infection segmentation network (ISNet) to perform segmentation. Also, we propose a scale uncertainty-aware aggregation method of prediction results. These methods enable the segmentation of infection regions of various sizes. ISNet has multiple encoder and a single decoder style structure. The use of the multiple encoders enables feature extraction from infection regions with a variation of CT values. Deep supervision is employed to improve the decoder's ability to decode the prediction result from feature value. ISNets having various receptive field sizes are trained and used to generate prediction maps from the CT volume. The scale uncertainty-aware prediction aggregation is applied to the multiple prediction maps to generate a final segmentation result considering uncertainty among the prediction results related to the receptive fields' size.

The contributions of this paper are (1) proposal of the ISNet with multiple encoders for feature extraction from infection regions that have a variation of CT values and (2) proposal of the scale uncertainty-aware aggregation method of prediction maps that are generated by segmentation models having a various size of receptive fields. These methods improve the segmentation accuracy of targets with significant variations in their intensity values and sizes.

2 Method

The proposed method segments infection regions from a chest CT volume of a COVID-19 patient. Set of patch volumes clipped from a CT volume is provided to ISNet. VRes and PSize define the size of the receptive field of ISNet. Change of the receptive field size causes variation on segmentation results (scale uncertainty). The scale uncertainty contains valuable information to refine segmentation results. We propose a scale uncertainty-aware aggregation process of segmentation results, which ISNets segment on various VRess and PSizes. The process generates a refined segmentation result.

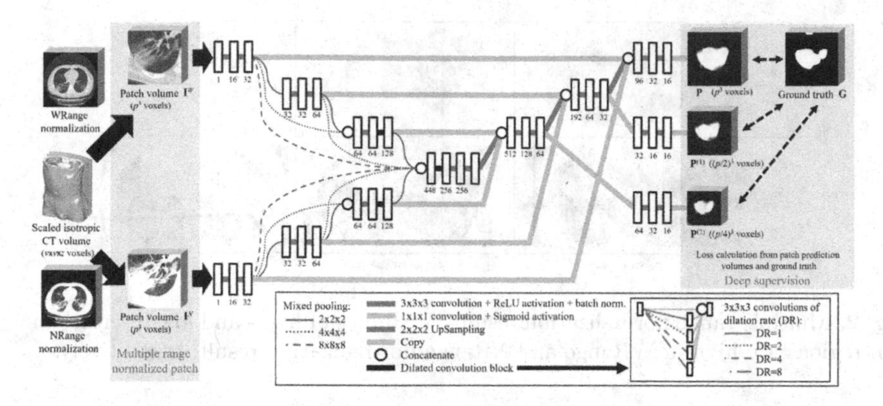

Fig. 1. Model structure of ISNet. Two encoders process patch volumes generated by different intensity normalizations. Encoders have dense pooling connections to bottleneck layer. Deep supervision is employed to evaluate subscale outputs.

2.1 Infection Region Segmentation by ISNet

Overview of Model: The structure of ISNet is shown in Fig. 1. ISNet has multiple encoders and a single decoder. Multiple volumes are generated from an input CT volume by applying CT value normalization by multiple value ranges to improve the segmentation accuracy of infection regions with various CT values. Patch volumes clipped from the volumes are input to ISNet. ISNet has multiple encoders corresponding to the multiple inputs to extract features in the CT value ranges selectively. The encoder has dense pooling connections [16] that prevent loss of spatial information by pooling layers. We employ deep supervision [17,18] in the decoder to improve its decoding performance from feature values.

Multiple Range Normalized Patch: An input CT volume is converted to a volume having an isotropic resolution in three dimensions. Then, the volume is scaled to $v \times v \times z$ voxels maintaining the aspect ratio. The number of voxels along the body axis z differs for each CT volume depending on its scanning range. CT values of infection regions distribute widely. CT values of consolidations range from -300 to 100 H.U.. GGO has a lower and broader range of CT values than the consolidations, ranging from -800 to 0 H.U.. CT value normalizations by multiple ranges are adequate for such a target. We apply CT value normalizations to the scaled CT volume using two CT value ranges, including; wide range (WRange): [-1000 H.U., 950 H.U.] and narrow range (NRange): [-1000 H.U., -400 H.U.]. Normalization results by the WRange are suitable to observe high-intensity infection regions, including consolidations and GGOs having high intensities. Normalization results by the NRange are suitable to observe GGOs having low intensities. Samples of normalization results are shown in Fig. 2. Two normalized volumes, including WRange volume and NRange volume, are generated from this process. We clip patch volumes from them at random positions. Patch volumes clipped from the WRange and NRange volumes are described as $\mathbf{I}^W_{v,p}, \mathbf{I}^N_{v,p} \in \mathbb{R}^{p \times p \times p}$, respectively.

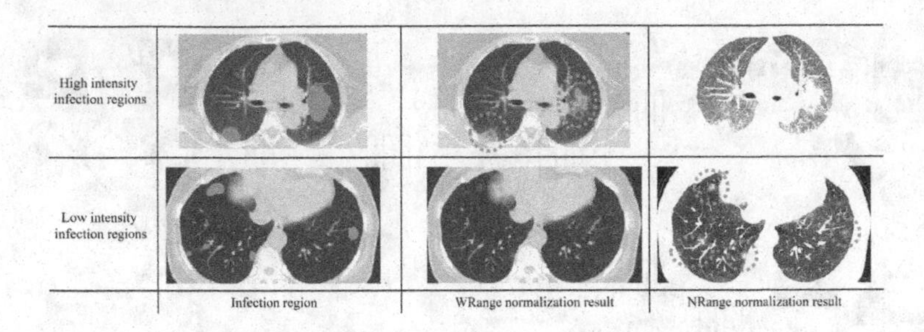

Fig. 2. Multiple range normalization results. Visibility of high- and low-intensity infection regions are high in WRange and NRange normalization result, respectively.

Multiple Encoding Paths: Inputs of the ISNet are patch volumes. We use two independent encoders to process $\mathbf{I}_{v,p}^{W}$ and $\mathbf{I}_{v,p}^{N}$. Feature values extracted by the encoders are concatenated at the bottleneck layer.

Pooling layers are commonly used in encoder, although it reduces spatial information in feature maps. The bottleneck layer connected after the encoder cannot receive enough spatial information. It causes segmentation results having incorrect boundaries. To reduce the loss of spatial information in the encoder, we adopt dense pooling connections [16] in the two encoders of ISNet. The dense pooling connections provide spatial information at each resolution in the encoder to the bottleneck layer. In the dense pooling connections, mixed poolings [16] are used instead of max poolings to reduce the loss of spatial information. Furthermore, we use dilated convolution [19] to utilize sparsely-distributed features in convolution operations. Dilated convolution block was implemented by connecting dilated convolutions of multiple dilation rates in parallel to obtain multiple-scales convolution results. Some dilated convolution blocks are inserted into ISNet.

Training: ISNet estimates a patch prediction volume $\mathbf{P}_{v,p} \in \mathbb{R}^{p\times p\times p}$ of infection regions from two input patch volumes $\mathbf{I}_{v,p}^{W}$ and $\mathbf{I}_{v,p}^{N}$. ISNet that performs estimation from input patch volumes ($p \times p \times p$ voxels) clipped from a volume ($v \times v \times z$ voxels) can be represented as a function $f_{v,p}$. Estimation of a patch prediction volume is formulated as

$$\mathbf{P}_{v,p} = f_{v,p}(\mathbf{I}_{v,p}^{W}, \mathbf{I}_{v,p}^{N}; \boldsymbol{\theta}_{v,p}), \tag{1}$$

where $\boldsymbol{\theta}_{v,p}$ is a parameter vector for infection region segmentation. The parameter vector is optimized in a supervised training process using CT volumes for training and their corresponding ground truth volumes $\mathbf{G} \in \{0,1\}^{p\times p\times p}$, whose elements 1 and 0 represent voxels in target or background regions. We employ deep supervision [17,18] for two subscales outputs. Their patch prediction volumes are $\mathbf{P}_{v,p}^{(1)}$ and $\mathbf{P}_{v,p}^{(2)}$. Their sizes are magnified to the same size as $\mathbf{P}_{v,p}$. The loss function to train ISNet is defined as

$$L = Dice(\mathbf{G}, \mathbf{P}_{v,p}) + \sum_{i=1}^{2} Dice(\mathbf{G}, \mathbf{P}_{v,p}^{(i)}), \tag{2}$$

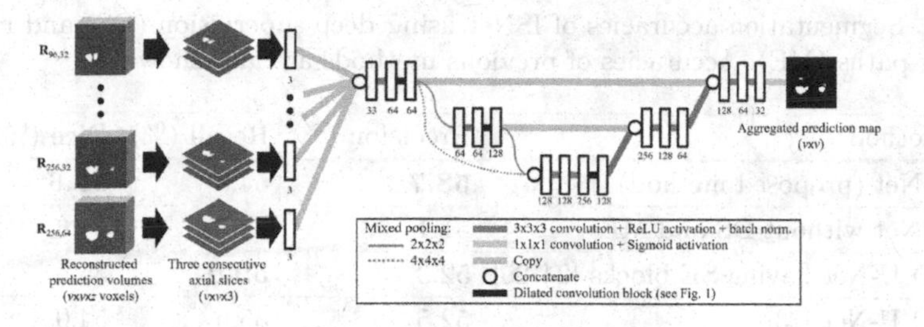

Fig. 3. Model structure of aggregation FCN. FCN processes axial slices obtained from prediction volumes. FCN outputs aggregation results from them.

where *Dice* is the dice loss between the ground truth volume and the patch prediction volumes.

Prediction: Patch volumes clipped from a CT volume for prediction are given to the trained ISNet $f_{v,p}$. The resulting patch prediction volumes are reconstructed as the same size as the CT volume. The reconstructed prediction volume is denoted as $\mathbf{R}_{v,p} \in \mathbb{R}^{v \times v \times z}$.

2.2 Scale Uncertainty-Aware Prediction Aggregation

The parameters v and p define the size of the receptive field of ISNet. The size of the receptive field of ISNet has a relationship to its segmentation accuracy. ISNets having various sizes of their receptive fields are trained and perform predictions, and we obtain multiple prediction volumes containing scale uncertainty from them. We utilize the scale uncertainty-aware aggregation method of the prediction volumes. An aggregation function is automatically trained based on each prediction volume's contribution to a segmentation result.

We train ISNets using training cases on multiple value settings of v and p. Using the ISNets, multiple reconstructed prediction volumes are generated from a CT volume for prediction. We perform aggregation of them using an aggregation FCN. The structure of the aggregation FCN is shown in Fig. 3. The FCN employs multiple 2D axial slice-based process. The FCN combines given prediction results by considering variation among them and how each prediction result contributes to generating a segmentation result. The FCN is trained using prediction volumes and their corresponding ground truth volumes of training cases. Generalized dice loss [20] is used to train the FCN. In a prediction process, outputs (on 2D axial slices) of the trained aggregation FCN are reconstructed to a 3D volume that has the same resolution and size as the original CT volume and thresholded to generate a segmentation result.

Table 1. Segmentation accuracies of ISNet using deep supervision (DS) and multiple encoding paths (ME). Accuracies of previous methods are also shown.

Method	Precision (%)	Recall (%)	Dice (%)
ISNet (proposed method)	**58.7**	54.6	**56.6**
ISNet without DS and ME	51.1	58.7	54.6
3D U-Net having SE blocks [24,25]	52.3	**59.0**	55.5
3D U-Net [23]	52.5	55.6	54.0

3 Experiments and Results

We evaluated segmentation accuracy of the proposed method. We used 199 chest CT volumes of COVID-19 patients provided by the Multi-national NIH Consortium for CT AI in COVID-19 [21] via the NCI TCIA public website [22]. The corresponding ground truth data of infection regions were also provided. We conducted three-fold cross-validations in our evaluations. Averaged precision, recall, and dice score among all CT volumes are used as the evaluation criteria. Methods were implemented using Keras 2.2.4 and TensorFlow 1.14.0. NVIDIA Tesla V100 GPU having 32 GB memory $\times 1$ was used to train and test the methods.

3.1 Ablation and Comparative Study of ISNet

We confirmed the segmentation performance of infection regions using ISNet. Techniques explained in 2.1 are used, including the deep supervision (DS) and the multiple encoding paths (ME). We confirmed the effectiveness of the technique in an ablation study. Dice scores of segmentation results obtained using ISNet and ISNet without DS and ME were calculated. ISNets were trained on parameter settings of $v = 192$, $p = 32$, minibatch size: 16, learning rate: 10^{-5}, and training epochs: 40. Adam was used as the optimization algorithm. The segmentation result was generated by applying thresholding to the prediction volume. Also, we compared dice scores of previous methods, including 3D U-Net [23] and 3D U-Net having squeeze-and-excitation (SE) blocks [24,25] with ISNet. These previous methods were applied to perform patch-based processes. The results are shown in Table 1. ISNet had a higher dice score than the previous methods. Also, the use of DS and ME contributed to improving the dice score.

3.2 Segmentation by Aggregation FCN

We applied the scale uncertainty-aware prediction aggregation to prediction volumes generated by ISNets. 11 prediction volumes generated by ISNets trained on parameter settings $(v, p) \in \{(256, 32), (256,64), (224,32), (224,64), (192,32), (192,64), (160,32), (160,64), (128,32), (128,64), (96, 32)\}$ were used. The aggregation FCN was trained on parameter settings of minibatch size: 4, learning rate: 10^{-3}, and training epochs: 5. Adam was used as the optimization algorithm.

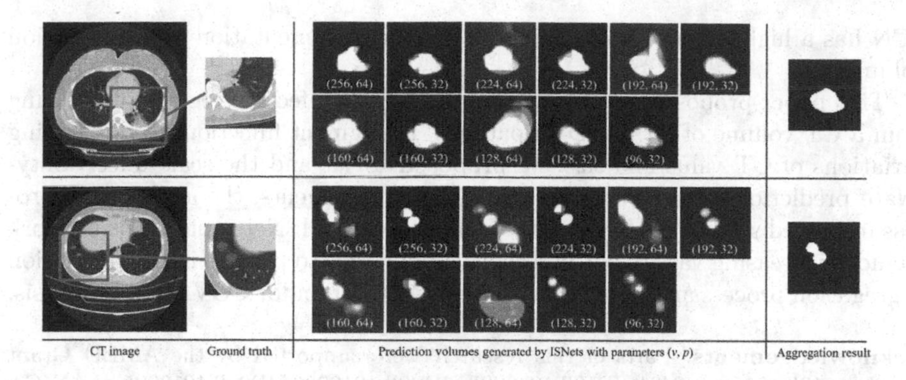

Fig. 4. Prediction volumes generated by ISNets with parameters (v, p) and aggregation results from them. While regions in prediction volumes have considerable variation, they are adequately aggregated.

Table 2. Segmentation accuracies of ISNets and aggregation result from them.

Method	Precision (%)	Recall (%)	Dice (%)
Aggregation FCN	**63.7**	**60.5**	**62.1**
Mean ± S.D. of 11 ISNets (before aggregation)	45.1 ± 12.3	52.6 ± 3.7	47.6 ± 8.1

Generated prediction volumes and aggregation results from them are shown in Fig. 4. Segmentation accuracies of ISNets and aggregation result are shown in Table 2. Accuracies were improved in all criteria by the aggregation.

4 Discussion and Conclusions

Segmentation of the lung infection region is difficult because it has significant variations in its CT values and shapes. We developed ISNet with the multiple range normalized patch processing paths and the scale uncertainty-aware prediction aggregation process to tackle infection segmentation having such difficulties. ISNet achieved higher segmentation accuracy than the previous methods, as shown in Table 1. Also, we confirmed the effectiveness of techniques, including deep supervision and multiple encoding paths in the ablation study. The scale uncertainty-aware prediction aggregation process improved the dice similarity score of the segmentation result. We used multiple prediction volumes generated by using multiple ISNets with various receptive fields' sizes. Because segmentation abilities and effective segmentation target sizes are different among the ISNets, an appropriate aggregation process of the prediction volumes can generate an accurate segmentation result. The aggregation FCN with trainable aggregation parameters was successfully built using training data. The evaluation result obtained in the cross-validation proved that the trained aggregation

FCN has a high generalization ability to perform segmentation from prediction volumes.

This paper proposed a segmentation method of infection regions in the lung from a CT volume of a COVID-19 patient. To segment infection regions having variations of CT value and size, we proposed ISNet and the scale uncertainty-aware prediction aggregation process. In our experiments, the aggregation process improved segmentation accuracy from individual ISNet results. Future work includes increasing variety of the receptive field sizes to process in the prediction aggregation process and development of a CAD system for COVID-19 diagnosis.

Acknowledgements. Parts of this research were supported by the AMED Grant Numbers 18lk1010028s0401, JP19lk1010036, JP20lk1010036, JP20lk1010025, the NICT Grant Number 222A03, the JST CREST Grant Number JPMJCR20D5, and the MEXT/JSPS KAKENHI Grant Numbers 26108006, 17H00867, 17K20099.

References

1. Coronavirus Update. https://www.worldometers.info/coronavirus/. Accessed 22 Aug 2021
2. Simpson, S., et al.: Radiological society of north America expert consensus document on reporting chest CT findings related to COVID-19: endorsed by the society of thoracic radiology, the American college of radiology, and RSNA. Radiol. Cardiothorac. Imaging **2**(2), e200152 (2020)
3. Ai, T., et al.: Correlation of chest CT and RT-PCR testing for coronavirus disease 2019 (COVID-19) in China: a report of 1014 cases. Radiology **296**(2), E32–E40 (2020)
4. Fan, D.-P., et al.: Inf-Net: automatic COVID-19 lung infection segmentation from CT images. IEEE Trans. Med. Imaging **39**(8), 2626–2637 (2020)
5. Wang, G., et al.: A noise-robust framework for automatic segmentation of COVID-19 pneumonia lesions from CT images. IEEE Trans. Med. Imaging **39**(8), 2653–2663 (2020)
6. Zheng, B., et al.: MSD-Net: multi-scale discriminative network for COVID-19 lung infection segmentation on CT. IEEE Access **8**, 185786–185795 (2020)
7. Mahmud, T., et al.: CovTANet: a hybrid tri-level attention based network for lesion segmentation, diagnosis, and severity prediction of COVID-19 chest CT scans. IEEE Trans. Ind. Inform. (Early Access) **39**(8), 2653–2663 (2020)
8. Yan, Q., et al.: COVID-19 Chest CT Image Segmentation - A Deep Convolutional Neural Network Solution. arXiv:2004.10987 (2020)
9. Roth, H.R., et al.: An application of cascaded 3D fully convolutional networks for medical image segmentation. Comput. Med. Imaging Graph. **66**, 90–99 (2018)
10. Oda, M., Roth, H.R., Kitasaka, T., et al.: Abdominal artery segmentation method from CT volumes using fully convolutional neural network. Int. J. Comput. Assist. Radiol. Surg. **14**, 2069–2081 (2019)
11. Kim, H., Jung, J., Kim, J., et al.: Abdominal multi-organ auto-segmentation using 3D-patch-based deep convolutional neural network. Sci. Rep. **10**, 6204 (2020)
12. Zhou, Y., Chang, H., Barner, K., Spellman, P., Parvin, B.: Classification of histology sections via multispectral convolutional sparse coding. IEEE Conf. Comput. Vis. Pattern Recogn. (CVPR), 3081–3088 (2014)

13. Xu, Y., Jia, Z., Ai, Y., Zhang, F., Lai, M., Chang, E.I.: Deep convolutional activation features for large scale brain tumor histopathology image classification and segmentation. IEEE Int. Conf. Acoust. Speech Signal Process. (ICASSP), 947–951 (2015)
14. Wang, D., Khosla, A., Gargeya, R., Irshad, H., Beck, A.H.: Deep Learning for Identifying Metastatic Breast Cancer. arXiv:1606.05718 (2016)
15. Tokunaga, H., Teramoto, Y., Yoshizawa, A., Bise, R.: Adaptive weighting multi-field-of-view CNN for semantic segmentation in pathology. IEEE/CVF Conf. Comput. Vis. Pattern Recogn. (CVPR), 12589–12598 (2019)
16. Playout, C., Duval, R., Cheriet, F.: A multitask learning architecture for simultaneous segmentation of bright and red lesions in fundus images. In: Frangi, A.F., Schnabel, J.A., Davatzikos, C., Alberola-López, C., Fichtinger, G. (eds.) Medical Image Computing and Computer Assisted Intervention – MICCAI 2018 , MICCAI 2018. LNCS, vol. 11071, pp. 101–108. Springer, Cham (2018). https://doi.org/10.1007/978-3-030-00934-2_12
17. Zeng, G., Yang, X., Li, J., Yu, L., Heng, P.-A., Zheng, G.: 3D U-net with multi-level deep supervision: fully automatic segmentation of proximal femur in 3D MR images. In: Wang, Q., Shi, Y., Suk, H.-I., Suzuki, K. (eds.) Machine Learning in Medical Imaging, MLMI 2017. LNCS, vol. 10541, pp. 274–282. Springer, Cham (2017). https://doi.org/10.1007/978-3-319-67389-9_32
18. Dou, Q., Yu, L., Chen, H., Jin, Y., Yang, X., Qin, J., Heng, P.-A.: 3D deeply supervised network for automated segmentation of volumetric medical images. Med. Image Anal. **41**, 40–54 (2017)
19. Yu, F., Koltun, V.: Multi-scale context aggregation by dilated convolutions. Int. Conf. Learn. Representations (ICLR) (2016)
20. Sudre, C.H., Li, W., Vercauteren, T., Ourselin, S., Cardoso, M.J.: Generalised dice overlap as a deep learning loss function for highly unbalanced segmentations. Int. Workshop Deep Learn. Med. Image Anal. (DLMIA) **10553**, 240–248 (2017)
21. An, P., et al.: CT Images in Covid-19 [Data set]. The Cancer Imaging Archive (2020)
22. Clark, K., et al.: The Cancer Imaging Archive (TCIA): maintaining and operating a public information repository. J. Digit. Imaging **26**(6), 1045–1057 (2013). https://doi.org/10.1007/s10278-013-9622-7
23. Çiçek, Ö., Abdulkadir, A., Lienkamp, S.S., Brox, T., Ronneberger, O.: 3D U-Net: learning dense volumetric segmentation from sparse annotation. In: Ourselin, S., Joskowicz, L., Sabuncu, M.R., Unal, G., Wells, W. (eds.) Medical Image Computing and Computer-Assisted Intervention – MICCAI 2016, MICCAI 2016. LNCS, vol. 9901, pp. 424–432. Springer, Cham (2016). https://doi.org/10.1007/978-3-319-46723-8_49
24. Roy, A.G., Navab, N., Wachinger, C.: Concurrent spatial and channel 'squeeze and excitation' in fully convolutional networks. In: Frangi, A.F., Schnabel, J.A., Davatzikos, C., Alberola-López, C., Fichtinger, G. (eds.) MICCAI 2018. LNCS, vol. 11070, pp. 421–429. Springer, Cham (2018). https://doi.org/10.1007/978-3-030-00928-1_48
25. Rundo, L., et al.: USE-Net: incorporating squeeze-and-excitation blocks into U-Net for prostate zonal segmentation of multi-institutional MRI datasets. Neurocomputing **365**, 31–43 (2019)

DCL

Multi-task Federated Learning for Heterogeneous Pancreas Segmentation

Chen Shen[1], Pochuan Wang[2], Holger R. Roth[3], Dong Yang[3], Daguang Xu[3], Masahiro Oda[1], Weichung Wang[2](✉), Chiou-Shann Fuh[2], Po-Ting Chen[4], Kao-Lang Liu[4], Wei-Chih Liao[4], and Kensaku Mori[1]

[1] Nagoya University, Nagoya, Japan
[2] National Taiwan University, Taipei, Taiwan
[3] NVIDIA Corporation, Santa Clara, USA
[4] National Taiwan University Hospital, Taipei, Taiwan

Abstract. Federated learning (FL) for medical image segmentation becomes more challenging in multi-task settings where clients might have different categories of labels represented in their data. For example, one client might have patient data with "healthy" pancreases only while datasets from other clients may contain cases with pancreatic tumors. The vanilla federated averaging algorithm makes it possible to obtain more generalizable deep learning-based segmentation models representing the training data from multiple institutions without centralizing datasets. However, it might be sub-optimal for the aforementioned multi-task scenarios. In this paper, we investigate heterogeneous optimization methods that show improvements for the automated segmentation of pancreas and pancreatic tumors in abdominal CT images with FL settings.

Keywords: Federated learning · Pancreas segmentation · Heterogeneous optimization

1 Introduction

Fully automated segmentation of the pancreas and pancreatic tumors from CT volumes is still challenging due to the low contrast and significant variations across subjects caused by different scanning protocols and patient populations. A great deal of progress has been made to improve the pancreas segmentation performance in the last decade with the rapid development of convolutional neural network (CNN) based approaches to this medical image segmentation task [9,11,12,20]. Still, highly accurate and generalizable pancreas and corresponding tumor segmentation models are encouraged as a prerequisite for computer-aided diagnosis (CAD) systems.

One limitation for pancreas segmentation using deep learning-based approaches is the lack of annotated training data. Distinct from natural images, collecting extensive training data from various resources may lead to multiple

C. Oyarzun Laura et al. (Eds.): CLIP/DCL/LL-COVID/PPML 2021, LNCS 12969, pp. 101–110, 2021.
https://doi.org/10.1007/978-3-030-90874-4_10

technical, legal, and privacy issues in healthcare applications. Federated learning (FL) is an innovative technique that enables to learn a deep learning-based model among distributed devices, i.e. clients, collaboratively without having to centralize the training data in one location [10].

FL techniques currently attract increasing attention in the medical image analysis field because the acquisition of annotated medical images from the real world is challenging and costly. A growing number of studies have been made to handle these difficulties by using the FL strategy. Furthermore, FL shows great effectiveness on segmentation tasks in abdominal organs [1,16], brain tumors [7,13] and COVID-19 image analysis [2,3,17,18] both in simulation and real-world FL applications.

One question remaining is on how to integrate best models trained on heterogeneous tasks among clients is difficult to determine. Most commonly, the Federated Averaging (FedAvg) is used to aggregate the model from each client and to update the global model on the server using a weighted sum where the weights are typically derived from the local training dataset sizes and kept constant during the training [10]. FedProx was introduced to handle data heterogeneity in FL by adding a regularization loss on the client that penalizes divergence from the current global model [6]. Techniques proposed for multi-task learning can be an alternative to tackle the heterogeneous statistical problems [15]. Dynamic task prioritization (DTP) specifies a prioritization for each task in multi-task learning based on the task-specific metrics [4]. Dynamic weight averaging (DWA) investigates a weight for each task calculated through the change of loss [8].

This work employs three public annotated datasets for pancreas segmentation to model three heterogeneous clients during FL. One dataset consists of pancreas and tumors, and the other two are consist of healthy pancreas cases. Our main contributions are as follows: 1) introduce the dynamic task prioritization for FL optimization; 2) investigate dynamic weight averaging aggregation method to re-weight the model from each client. 3) compare the effect of our improvements with FedAvg and FedProx on the pancreas and tumor segmentation task.

2 Methods

A standard FL system consists of a server and several clients. In a new FL round, each client receives the global model from the server and fine-tunes it on their local dataset. Then the client only shares a weight update with the server after the local training. The server is designed to receive the model updates from the specified minimum of clients and aggregates the updates based on the aggregation weight of each client. It then updates the global model with the aggregated updates and distributes the updated global model for the next round of FL training. An illustration of the FL system is shown in Fig. 1. The standard FL tries to minimize

$$\mathcal{L} = min \sum_{k}^{K} \eta_k \mathcal{L}_k, \tag{1}$$

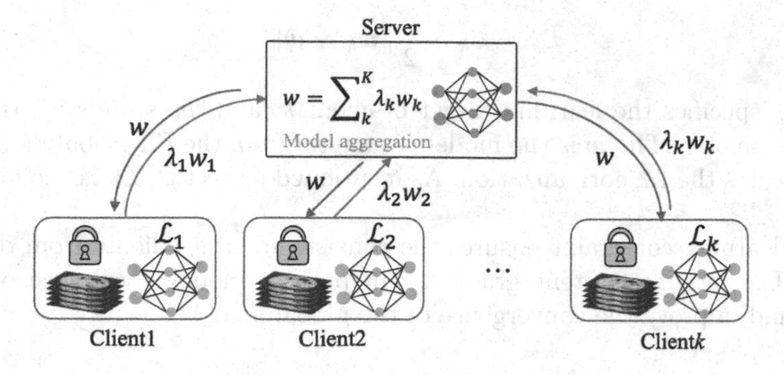

Fig. 1. An illustration of federated learning in medical imaging. The server only receives model updates and the training data stays on the client sites privately.

where the k-th client tries to optimize the local loss function $\mathcal{L}_k$. The total number of clients is K and the proportion that each client contributes to the global model update is $\eta_k \geq 0$, where $\sum_k^K \eta_k = 1$. In this work, our aim is to investigate a multi-task FL optimization method for heterogeneous pancreas segmentation where clients exhibit different types of images and labels in their data.

In this section, we first revisit the FedAvg [10] and FedProx [6] methods which are widely used in FL tasks. Then, we adapt two optimization methods from the multi-task learning literature to the FL setting: *dynamic task prioritization* and *dynamic weight averaging*.

2.1 FedAvg

In standard FedAvg, to reduce the heavy communication cost and to handle dropping clients, only a subset K clients are used, instead of total N clients to update the global model in each round. Here, we have $K << N$. The weight η_k of client $k \in K$ is a constant number which can be calculated by

$$\eta_k = \frac{n_k}{n}, \tag{2}$$

where n_k is the number of local training data in client k. The total number of training data in all clients can be derived from $n = \sum_k^K n_k$. In FedAvg, the client with larger local training data contributes more to the updated global model.

2.2 FedProx

FedProx is an improved federated optimization algorithm for learning from distributed heterogeneous datasets [6]. The FedProx algorithm is an extension of the standard FedAvg scheme. The FedProx algorithm adopted the aggregation scheme and added another learning constraint for each client, namely a regularization The regularization term can help the local client model to stay close to server model. The local client try to minimize

$$\hat{\mathcal{L}}_k = \mathcal{L}_k + \frac{\mu}{2} \|\boldsymbol{w}_k - \boldsymbol{w}\|^2, \quad (3)$$

where $\hat{\mathcal{L}}_k$ specifies the learning target of client k, and $\boldsymbol{w}_k$ stands for the local model parameters.The $\boldsymbol{w}$ is the model parameter from the FL global model, and $\|\cdot\|^2$ indicates the L2 normalization. As mentioned in Sect. 2, $\mathcal{L}_k$ is the local loss function.

This learning constraint ensures the consistency of gradients from different clients. The more consistent gradient can prevent model divergence of client models and improve the convergence of the global model.

2.3 Dynamic Task Prioritization

Dynamic task prioritization (DTP) for multi-task learning adjusts the weights between different tasks by estimating the key performance index (KPI) κ. κ is a monotone increasing function ranged from 0 to 1; the larger value of κ stands for better performance of the specific task. DTP concentrates on challenging tasks by increasing corresponding weights and lowering the weights of easier tasks. We generalized the DTP for federated learning by considering each client as a different task. In this work, we define the KPI of client k as

$$\kappa_{k,i} = d_{k,i}^r, \quad (4)$$

where the value of $\kappa_{k,i}$ is the power r of the Dice score $d_{k,i}$ of the i-th training batch. In multi-class segmentation we use the average Dice score of all classes except background. The final weight $W_{k,i}$ applies to the loss is defined as

$$W_{k,i} = -(1 - \bar{\kappa_{k,i}})^\gamma \log \bar{\kappa_{k,i}}. \quad (5)$$

To stabilize the weights between each batch, an exponential average $\bar{\kappa_{k,i}}$ is used:

$$\bar{\kappa_{k,i}} = (1 - \alpha)\, \kappa_{k,i} + \alpha \kappa_{k,\bar{i-1}}. \quad (6)$$

Here, α is a number between 0 and 1 and γ is a tunable hyperparameter.

2.4 Dynamic Weight Averaging

In dynamic weight averaging (DWA), we try to optimize the FL procedure by focusing on server model aggregation instead of applying a constraint on loss function. This method is inspired by optimization approaches from classical multi-task learning tasks [8]. In FL, finding a suitable balance to aggregate the model updates from heterogeneous clients is challenging. However, to specify the proper weight requires a large number of experiments and priority knowledge. In DWA, we investigate a method that defines the client weights on each round automatically, The server learns to weigh each client based on the variation of loss values from the previous round. The weight of client k in round r can be define as

$$\lambda_{k,r} = \frac{\xi \exp(\rho_{k,r-1}/T)}{\sum_{i=1}^{K} \exp(\rho_{i,r-1}/T)}, \quad (7)$$

where $\rho_{k,r-1} \in (0, +\infty)$ represents the dynamic proportion of the loss value $\mathcal{L}$ changes in client k from the round before previous round $r-2$ to the previous round $r-1$, which can be defined as $\rho_{k,r-1} = \mathcal{L}_{k,r-1}/\mathcal{L}_{k,r-2}$. To control the effeteness of dynamic proportion, T is defined as similar in the MTL in [8]. When $T \rightarrow +\infty$, the weight of each client is tend to be equally $\rho_k \rightarrow 1$. We introduce $\xi \in \mathbb{N}$ to adjust the impact of weights in DWA. Different from the way to calculate loss value $\mathcal{L}_{k,r}$ in [8], we average the local loss value of each iteration in one round, which can be defined as

$$\mathcal{L}_{k,r} = \frac{1}{J}\sum_{j=1}^{J}\mathcal{L}_{k,j}, \tag{8}$$

where j is the local iteration number within the total J iterations. The average operation will make the loss value of each round more stable. For the first round (when $r = 1$), we initialize the $\mathcal{L}_{k,r-1} = 1, \mathcal{L}_{k,r-2} = 1$ so that we can have the $\rho_{k,1} = 1$ after the first round.

3 Experiments and Results

3.1 Datasets

The experiment is conducted with one federated server for model aggregation and three clients for training. The server does not own any validation data and only aggregates the client's model parameters. Each client owns a dataset from a different source. The first dataset is the Pancreas-CT from The National Institutes of Health Clinical Center (TCIA)[1] [12]. This dataset contains 82 abdominal contrast-enhanced CT scans with manual segmentation labels for the healthy pancreas. The second dataset is the Task07 pancreas from the Medical Segmentation Decathlon challenge[2] (MSD) [14]. This dataset contains 281 portal venous phase CT scans with manual labels for the pancreas and pancreatic tumors (intraductal mucinous neoplasms, pancreatic neuroendocrine tumors, or pancreatic ductal adenocarcinoma). The third dataset is from the MICCAI Multi-Atlas Labeling Beyond the Cranial Vault challenge (Synapse)[3] [5]. This dataset contains 30 portal venous contrast phase CT scans with manual labels for 13 abdominal organs includes the pancreas. We only keep the pancreas labels for the third dataset and discard the labels for the other 12 organs. We randomly shuffled the three datasets separately and split them into training, validation, and testing sets with the ratio of 60%, 20%, and 20%. Among the total 231 training cases, 165 cases have both pancreas and pancreatic tumor labels.

---

[1] https://wiki.cancerimagingarchive.net/display/Public/Pancreas-CT.
[2] http://medicaldecathlon.com.
[3] https://www.synapse.org/#!Synapse:syn3193805/wiki/217785.

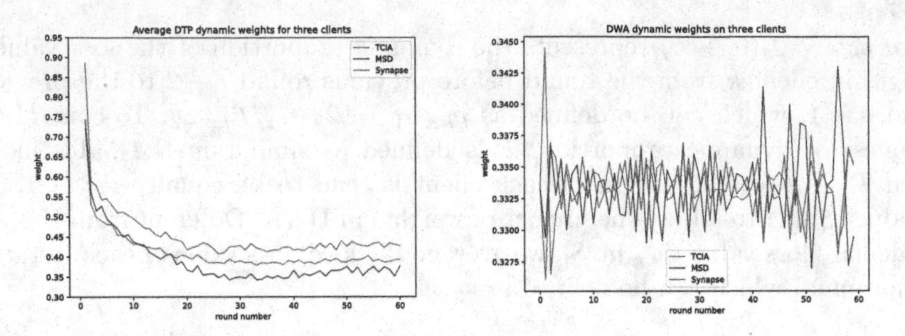

Fig. 2. Each client's weight is chosen by DTP and DWA method every round.

3.2 Experimental Details

We use NVIDIA Clara Train SDK 3.1[4] as the federated learning framework. During the experimentation, the server and associated clients are physically on the same machine and running in individual Docker containers. The server has no access to GPU, and each client has one V100 32 GB GPU. We run the experiments on two machines, the first one is a DGX-Station with 20 CPU cores, 256 GB system memory, and 4 V100 32 GB GPUs, and the second one is a DGX-1 with 40 CPU cores, 512 GB system memory, and 8 V100 GPUs. All CT volumes were resampled to isotropic spacing with $1.0 \times 1.0 \times 1.0\ \text{mm}^3$. To ensure the CT volumes were in the same orientation, we arranged the voxel axes as close as possible to RAS+ orientation. The Hounsfield unit (HU) intensity in the range [−200, 250] HU were rescaled and clipped into [−1, 1]. We used a network found by coarse-to-fine architecture search (C2FNAS) [19] using a TensorFlow implementation in all experiments. The training loss function is the sum of Dice loss and cross entropy. The Adam optimizer with cosine annealing learning rate scheduler was adopted with the initial learning rate 5×10^{-4}. The input patch size of our network is $96 \times 96 \times 96$. The total round number of FL was 60 with local epoch number of 10. The minimum client number was 3. Despite running FL in simulation on public datasets, we employed a percentile sharing protocol as a privacy-preserving measure [7]. We only share 25% of the model updates with the largest absolute values to ensure that our approach could be employed in a real-world setting.

3.3 Results

Our experimental results include the standalone training model on each dataset (TCIA local, MSD local, and Synapse local) and the FL global best model (determined using the average client validation scores during each FL round) for FedAvg, FedProx, DTP, and DWA. Table 1 compares the Dice score across all experiments with different hyperparameter settings. For standalone training model: TCIA local, MSD local, and Synapse local, the performance on other

[4] https://docs.nvidia.com/clara/clara-train-archive/3.1/index.html.

Table 1. Comparison of Dice Score for the pancreas and tumor segmentation on *local models* which trained from scratch with single datasets (TCIA, MSD and Synapse); and on *FL server best global model* with FedAvg, FedProx, DTP and DWA optimization. Best scores are shown in **bold**. *Italic* scores indicate the local models performance on its own test data. Non-italic numbers show the lack of generalizability of local models evaluated on other clients' test data.

	TCIA	MSD		Synapse	All
	Pancreas	Pancras	Tumor	Pancreas	Avg
TCIA local ($N_{train} = 48$)	*79.4%*	71.8%	0.0%	5.0%	40.1%
MSD local ($N_{train} = 165$)	61.9%	*77.8%*	*31.1%*	4.4%	40.3%
Synapse local ($N_{train} = 18$)	9.8%	0.4%	0.0%	*61.1%*	23.7%
FedAvg [10]	80.6%	**75.1%**	**20.2%**	42.6%	56.9%
FedProx [6]	80.6%	75.0%	19.5%	47.6%	58.5%
DTP ($\gamma = 1, \alpha = 0.9, r = 1$)	64.1%	54.8%	14.4%	34.5%	44.4%
DTP ($\gamma = 2, \alpha = 0.9, r = 1$)	46.0%	57.3%	12.4%	27.6%	36.1%
DTP ($\gamma = 1, \alpha = 0.5, r = 1$)	64.4%	57.2%	12.4%	39.2%	46.1%
DTP ($\gamma = 1, \alpha = 0.5, r = 2$)	64.3%	55.7%	12.4%	39.3%	45.9%
DWA ($T = 1$)	65.3%	59.8%	9.5%	49.8%	49.9%
DWA ($T = 1.5$)	76.1%	71.7%	13.7%	53.0%	57.3%
DWA ($T = 2$)	78.2%	72.4%	6.8%	56.2%	58.0%
DWA ($T = 2, \xi = 2$)	**80.9%**	73.4%	13.9%	**59.6%**	**61.4%**
DWA ($T = 2, \xi = 3$)	68.7%	59.8%	7.7%	39.3%	47.3%

datasets is quite unsatisfactory. FL global models have markedly better generalizability than standalone models. Both DTP and DWA methods rely on precise hyperparameter settings. For FedAvg, the Dice score on the MSD dataset is highest; however, the performance is not ideal on the Synapse dataset. The average Dice score with FedProx improves over FedAvg. The TCIA dataset and Synapse dataset have the highest Dice score with DWA ($T = 2, \xi = 2$). The average Dice score with DWA is 4.5% and 2.9% higher than FedAvg and FedProx.

Axial visualizations of the segmentation results are shown in Fig. 3. The performance of using FedAvg and FedProx is close: the segmentation results on TCIA and MSD data are acceptable but not ideal on relatively small Synapse data. In contrast, the DWA method shows a more stable performance on three different datasets. We visualize the dynamically chosen weights by the DTP, and DWA approaches in Fig. 2.

4 Discussion

As shown in Table 1, the FedAvg is the standard federated learning baseline to compare with other methods. Three local models are standalone-training results

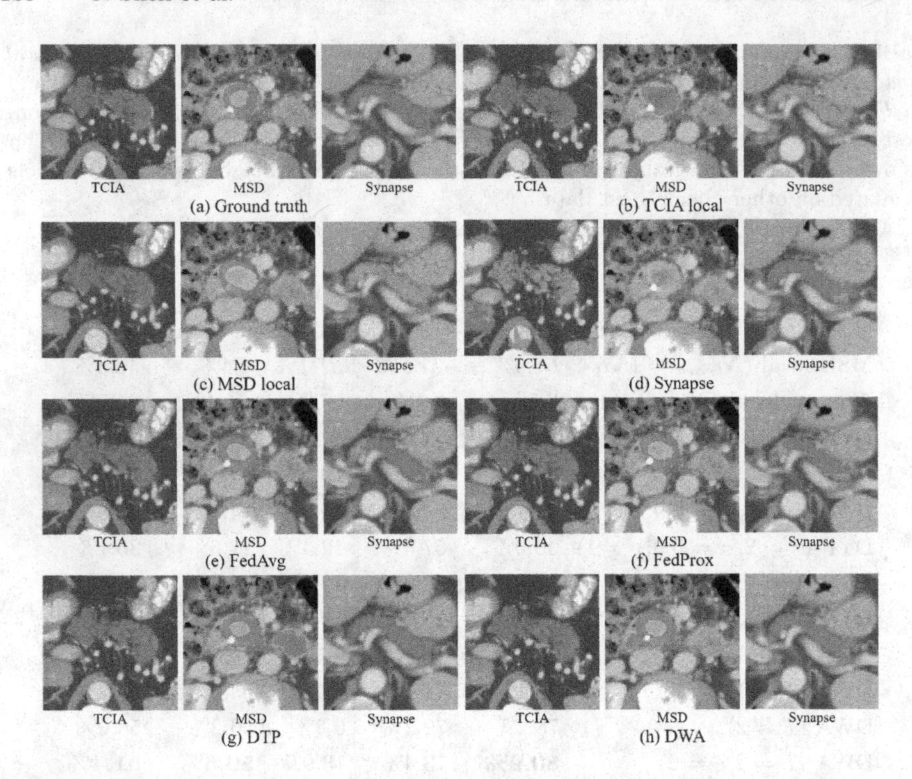

Fig. 3. Examples of pancreas and tumor segmentation on (b) TCIA local, (c) MSD local, (d) Synapse local, and on FL server global models with (e) FedAvg [10], (f) FedProx [6], (g) DTP and (h) DWA of TCIA, MSD and Synapse dataset, respectively.

for their corresponding datasets. The global model resulting from FedAvg performs well in the TCIA dataset, and the MSD pancreas compares to local models. Moreover, for the MSD tumor and Synapse dataset, although the performance is not as high as corresponding local models, there is still a significant improvement to other local models, indicating the improved generalizability of the global models. The FedProx model shows similar performance as the FedAvg model. In the MSD dataset, the average Dice score of the pancreas and tumor is slightly lower than the result of the FedAvg model. However, the average Dice score of the Synapse dataset is significantly higher than the result of the FedAvg model. The performance of DTP models is generally lower than the FedAvg baseline and DWA results. Nonetheless, in most settings, DTP models still outperform the local models. Furthermore, compared to the DWA results, the performance on MSD tumors is more consistent. Both FedAvg and FedProx are commonly used in FL, and our experiments suggest that both methods already provide a strong baseline performance even on heterogeneous datasets.

In DTP, the dynamic prioritization weight focuses on the most challenging tasks by adjusting the magnitude of the loss. However, each client only calculates the prioritization weights using local batch data. The lack of a global perspective of the training can therefore limit the performance of DTP. Also, DTP scales the magnitude of the loss, disrupting the optimization and increasing the need for further hyperparameter tuning. In contrast, with most DWA configurations, the Synapse dataset's performance is markedly higher than the FedAvg baseline. The results show that DWA can outperform both FedAvg and FedProx on average with properly selected hyperparameters.

5 Conclusion

In this work, we investigated two multi-task optimization methods for FL in medical imaging with heterogeneous datasets: DTP and DWA. The application of both methods was inspired by the analogy and similarity between FL and multi-task learning. We evaluated each method within an FL framework and compared the global model performance with FedAvg and FedProx. The Dice of DTP is lower than FedAvg and FedProx, likely because of limited manual tuning. However, the global model from DTP still outperforms the local models. DWA model aggregation method shows significant improvement, especially on the Synapse client whose training data is relatively smaller than other two clients.

Acknowledgement. Parts of this research was supported by the MEXT/JSPS KAKENHI (894030, 17H00867).

References

1. Czeizler, E., et al.: Using federated data sources and varian learning portal framework to train a neural network model for automatic organ segmentation. Physica Med. **72**, 39–45 (2020)
2. Dou, Q., et al.: Federated deep learning for detecting COVID-19 lung abnormalities in CT: a privacy-preserving multinational validation study. Npj Digit. Med. **4**(1), 60 (2021). https://doi.org/10.1038/s41746-021-00431-6
3. Flores, M., et al.: Federated learning used for predicting outcomes in SARS-COV-2 patients (2021). https://doi.org/10.21203/rs.3.rs-126892/v1
4. Guo, M., Haque, A., Huang, D.A., Yeung, S., Fei-Fei, L.: Dynamic task prioritization for multitask learning. In: Proceedings of the European Conference on Computer Vision (ECCV), pp. 270–287 (2018)
5. Landmanm, B., et al.: 2015 MICCAI multi-atlas labeling beyond the cranial vault - workshop and challenge (2015). https://doi.org/10.7303/syn3193805
6. Li, T., Sahu, A.K., Zaheer, M., Sanjabi, M., Talwalkar, A., Smith, V.: Federated optimization in heterogeneous networks. arXiv preprint arXiv:1812.06127 (2018)
7. Li, W., et al.: Privacy-preserving federated brain tumour segmentation. In: Suk, H.-I., Liu, M., Yan, P., Lian, C. (eds.) MLMI 2019. LNCS, vol. 11861, pp. 133–141. Springer, Cham (2019). https://doi.org/10.1007/978-3-030-32692-0_16
8. Liu, S., Johns, E., Davison, A.J.: End-to-end multi-task learning with attention. In: Proceedings of the IEEE/CVF Conference on Computer Vision and Pattern Recognition (CVPR) (2019)

9. Man, Y., Huang, Y., Feng, J., Li, X., Wu, F.: Deep Q learning driven CT pancreas segmentation with geometry-aware U-net. IEEE Trans. Med. Imaging **38**(8), 1971–1980 (2019)
10. McMahan, H.B., Moore, E., Ramage, D., Hampson, S., y Arcas, B.A.: Communication-efficient learning of deep networks from decentralized data. In: AISTATS (2017)
11. Oktay, O., et al.: Attention U-net: learning where to look for the pancreas. arXiv preprint arXiv:1804.03999 (2018)
12. Roth, H.R., et al.: DeepOrgan: multi-level deep convolutional networks for automated pancreas segmentation. In: Navab, N., Hornegger, J., Wells, W.M., Frangi, A.F. (eds.) MICCAI 2015. LNCS, vol. 9349, pp. 556–564. Springer, Cham (2015). https://doi.org/10.1007/978-3-319-24553-9_68
13. Sheller, M.J., Reina, G.A., Edwards, B., Martin, J., Bakas, S.: Multi-institutional deep learning modeling without sharing patient data: a feasibility study on brain tumor segmentation. In: Crimi, A., Bakas, S., Kuijf, H., Keyvan, F., Reyes, M., van Walsum, T. (eds.) BrainLes 2018. LNCS, vol. 11383, pp. 92–104. Springer, Cham (2019). https://doi.org/10.1007/978-3-030-11723-8_9
14. Simpson, A.L., et al.: A large annotated medical image dataset for the development and evaluation of segmentation algorithms. arXiv preprint arXiv:1902.09063 (2019)
15. Smith, V., Chiang, C.K., Sanjabi, M., Talwalkar, A.: Federated multi-task learning. In: Proceedings of the 31st International Conference on Neural Information Processing Systems, pp. 4427–4437 (2017)
16. Wang, P., et al.: Automated pancreas segmentation using multi-institutional collaborative deep learning. In: Albarqouni, S. (ed.) DART/DCL -2020. LNCS, vol. 12444, pp. 192–200. Springer, Cham (2020). https://doi.org/10.1007/978-3-030-60548-3_19
17. Xia, Y., et al.: Auto-FedAvg: learnable federated averaging for multi-institutional medical image segmentation (2021)
18. Yang, D., et al.: Federated semi-supervised learning for COVID region segmentation in chest CT using multi-national data from china, Italy, Japan. Med. Image Anal. **70**, 101992 (2021)
19. Yu, Q., et al.: C2FNAS: coarse-to-Fine neural architecture search for 3D medical image segmentation, December 2019
20. Zhou, Y., Xie, L., Shen, W., Wang, Y., Fishman, E.K., Yuille, A.L.: A fixed-point model for pancreas segmentation in abdominal CT scans. In: Descoteaux, M., Maier-Hein, L., Franz, A., Jannin, P., Collins, D.L., Duchesne, S. (eds.) MICCAI 2017. LNCS, vol. 10433, pp. 693–701. Springer, Cham (2017). https://doi.org/10.1007/978-3-319-66182-7_79

Federated Learning in the Cloud for Analysis of Medical Images - Experience with Open Source Frameworks

Przemysław Jabłecki[1], Filip Ślazyk[1], and Maciej Malawski[1,2](✉)

[1] Institute of Computer Science, AGH University of Science and Technology, Krakow, Poland

[2] Sano Centre for Computational Medicine, Krakow, Poland

malawski@agh.edu.pl

https://sano.science

Abstract. Federated Learning (FL) is a novel technique that allows for performing the training of a global model without sharing data between entities. This research focused on the analysis of existing solutions for Federated Learning in the context of medical image classification. Selected frameworks: TensorFlow Federated, PySyft and Flower were tested and their usability was assessed. Additionally, experiments on classification of X-ray lung images with the use of the Flower framework were performed in a fully distributed setting using Google Cloud Platform.

1 Introduction

A wide range of algorithms for the analysis of medical images requires a massive amount of data. Nonetheless, entities such as hospitals are usually forbidden to share any patients' information to prevent privacy breaches and to conform with data protection regulations. Hence, it is essential to find a way to utilise such privacy-sensitive data without the need to gather them in a centralized system. The solution to this problem is *Federated Learning* [10], which takes advantage of multiple clients collaboratively training a shared global model.

The main focus of this research was to train neural networks to classify the following diseases of the respiratory system: COVID-19 and pneumonia, following the Federated Learning approach in a fully distributed way to assess real-life usability and robustness of such a technique. By a *fully distributed* scenario we mean running the analysis on multiple separate client machines, connected to a central server over the network. As there are several Open Source frameworks available, we intended to evaluate their usability for such a scenario. Moreover, we evaluated the performance of the models under IID (independent and identically distributed data) and NonIID conditions.

2 Related Work

Federated learning is becoming an important technique for the analysis of medical data [13], in particular for medical imaging [8]. A notable example of such

C. Oyarzun Laura et al. (Eds.): CLIP/DCL/LL-COVID/PPML 2021, LNCS 12969, pp. 111–119, 2021.
https://doi.org/10.1007/978-3-030-90874-4_11

study is the analysis of MRI data for prostate cancer diagnosis, using data from 3 academic institutions [12]. The FL model in the study showed significantly better performance and generalizability to the models trained at single institutions. An example of this study, where FL was run in a distributed environment was presented in [14], where Fed-BioMed framework was used to run the analysis on a brain imaging dataset from 4 centers using a variational autoencoder model.

An attempt, similar to ours, to run FL analysis using chest X-Ray image data for COVID-19 was made by Liu et al. [9]. Nevertheless, that research was not conducted in a fully distributed manner as only one GPU (NVIDIA Tesla V100) was used [9]. Another recent study shows the results of a multicenter FL analysis of the COVID-19 CT scans, where data from 3 hospitals from Hong Kong was used for the model training, and the data from Germany and China were used for validation [7]. Again, the study was not performed in a fully distributed environment, as all data was processed in a single machine (see: https://github.com/med-air/FL-COVID).

3 Dataset Used in Evaluation

For our evaluation, we used the public dataset available on Kaggle under the name "COVID-19 Radiography Database" [1]. To obtain a balanced dataset, we undersampled the original data. After performing that phase, the dataset consisted of samples belonging to three classes: "pneumonia", "covid19" and "normal". Each class had 2690 samples. The dataset was split into 3 parts: train, validation and test (split ratio - train: 0.6, validation: 0.2, test: 0.2). The images had the resolution of 224×224 px. Sample images are presented in Fig. 1.

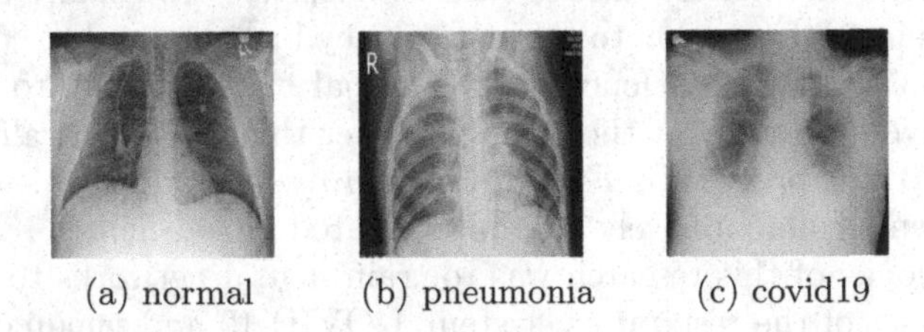

(a) normal (b) pneumonia (c) covid19

Fig. 1. Sample images from each class

4 Overview of Available Open Source Frameworks for FL

To conduct research on usability of Federated Learning, we analysed and compared the following frameworks: TensorFlow Federated, PySyft and Flower.

4.1 TensorFlow Federated

TensorFlow Federated [5] is an open-source framework developed by Google to support computations on decentralized data. One of the advantages of this solution is that it incorporates TensorFlow's API which results in a shallow learning curve for data scientists familiarized with the platform. The drawback of this project is the fact that it allows only for local simulations. For that reason, we used this framework only for test computations on a single machine.

4.2 PySyft

PySyft [4] is a library that focuses on secure and decentralized Deep Learning [11]. It utilises Differential Privacy and Encrypted Computation. It has been developed by OpenMined open-source community. PySyft works as a hook to the PyTorch framework. What is more, the library allows for experimenting with two types of Federated Learning: model-centric and data-centric. The important part of PySyft is PyGrid - a piece of software working as a central server for training models at a large scale [3]. Having analysed the capabilities of the library, we decided to move forward although it had minor issues caused by the rapid development. In our evaluation, we deployed a central server on Google Cloud Platform and performed simulations with simple CNNs to classify MNIST dataset utilising Jupyter Notebooks as clients. However, due to problems with the current implementation, we did not manage to run a more complex analysis.

4.3 Flower

Flower is a recent Federated Learning framework. The main motivation behind it was to build software free from the issues of competing solutions [6] - to allow for research on FL and deliver production-ready environments. The foremost feature of Flower is its framework-agnostic implementation. What is more, it provides high-level abstraction and a convenient API. Flower uses gRPC library [2] for communication between the client and the server. The results presented in the following sections were obtained with the use of this framework.

5 Experiment Setup

The goal of the experiment was to evaluate the performance of federated training of our medical image classification model in a fully distributed environment with 1 server and the number of clients between 1 and 4 (Fig. 2). The central server was deployed as a single Compute Engine unit (c2-standard-4) on the Google Cloud Platform. The clients were run with the use of NVIDIA Tesla K80 on Google Collab instances. Fedarated Learning abstraction was implemented with the use of Flower framework. What is more, Terraform was used to obtain automation and high-level flexibility of deployment.

We analysed two neural network architectures (EfficientNetB0 and ResNet50), and evaluated the impact of the number of local epochs (LE) and

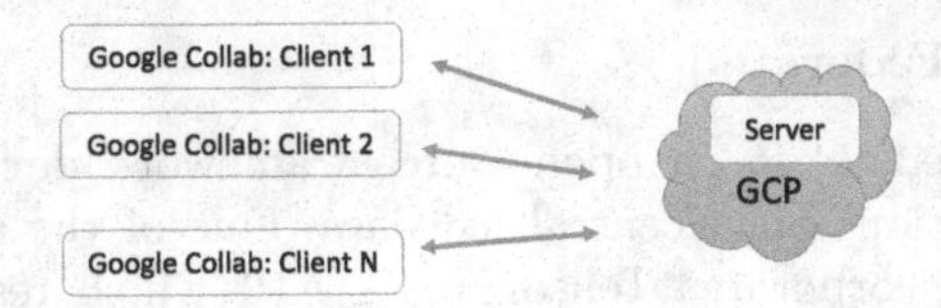

Fig. 2. Distributed experiment setup on google cloud platform and collab

client count (CC) on the overall performance. The neural network models available in Keras library in TensorFlow framework were utilised. Federated Learning was conducted with *FedAvg* averaging strategy originally proposed in [10]. During each round, 0.5 of the total number of clients were selected for training.

For each model, two data distribution strategies were tested. The first one is called 'IID' which stands for independent and identically distributed data among each client (the same number of samples in each class on every client). Always the whole dataset was partitioned among all clients. 'NonIID' strategy was defined as follows, to obtain the "worst" possible NonIID distribution, where some clients had samples belonging to only one class:

1. Order all available image samples in a list according to their class, such that images of the same class are grouped together.
2. Split the list into k equal length splits, where k denotes the number of clients.
3. Assign splits to the clients according to their individual *id*s (clients were numbered from 0 to $k-1$).

6 Results

We present the results of the experiments, grouped by the architecture of the model and the data distribution scenarios.

6.1 Results for EfficientNetB0 Architecture

NonIID: Results are presented in Fig. 3. In the case of 3 clients, the maximal accuracy achieved over 20 rounds was 0.367. For 4 clients maximum accuracy was 0.551. In both cases, the accuracy and loss changes were not monotonic.

IID: In the case of IID strategy, we have found that increasing the number of local epochs resulted in a faster convergence of the training (a similar observation was made in [10] in the case of image classification). Results for 3 and 4 clients were presented in Fig. 4. Maximum accuracy of each training in that setup was between 91%-93%. The training converged around the 6th round for all tested combinations of local epochs count and both for 3 and 4 clients.

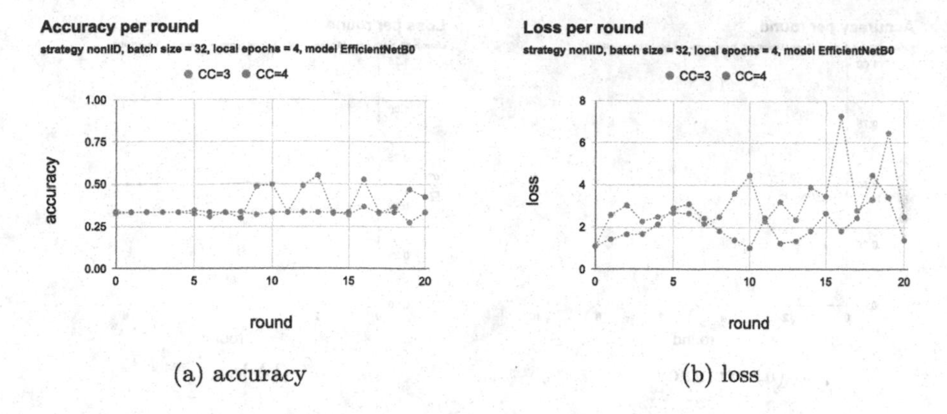

(a) accuracy (b) loss

Fig. 3. Model EfficientNetB0, strategy NonIID, 3 and 4 clients

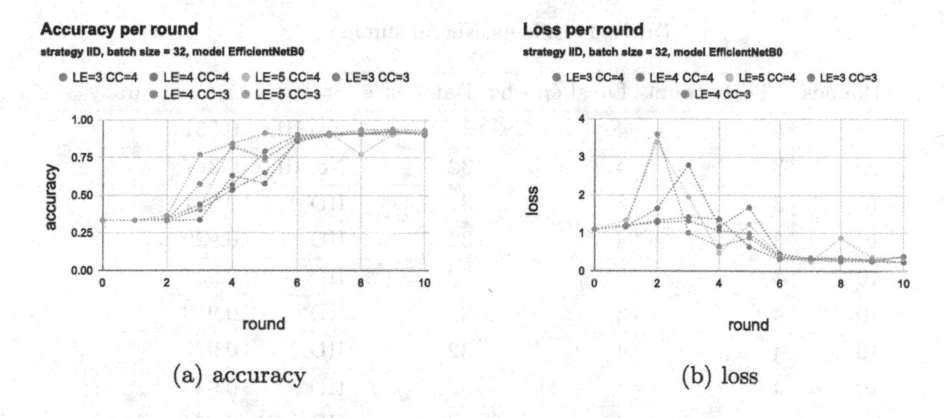

(a) accuracy (b) loss

Fig. 4. Model EfficientNetB0, strategy IID, 3 and 4 clients

IID 1 Client: For reference, we have tested the training performed on a single client (Fig. 5). In that case, the training converged much faster (around the 2nd and 3rd round) than in the case of 3 and 4 clients. Maximum achieved accuracy was 0.944, which is the highest value of all measured accuracies in the case of EfficientNetB0.

All training results for EfficientNetB0 model were summarized in Table 1.

6.2 Results for ResNet50 Architecture

NonIID: In the case of NonIID distribution with ResNet50 model, the achieved results were comparable to those achieved for EfficientNetB0 architecture. Maximum accuracy for 4 clients was 0.560 and 0.390 for 3 clients (in Fig. 6).

IID: For IID distribution, the training converged slower than in the case of EfficientNetB0 model, only to achieve top accuracy between 8th - 10th round, both in the case of 3 and 4 clients (Fig. 7). It was observed that in the case of

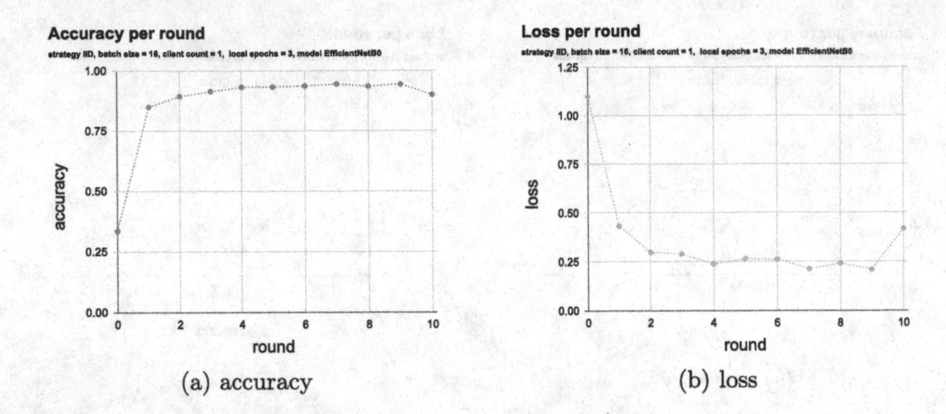

(a) accuracy (b) loss

Fig. 5. Model EfficientNetB0, strategy IID, 1 client

Table 1. EfficientNetB0 summary

Rounds	Clients count	Local epochs	Batch size	Strategy	Max accuracy
20	4	4	32	NonIID	0.551
20	3	4	32	NonIID	0.367
10	3	3	32	IID	0.921
10	3	4	32	IID	0.929
10	3	5	32	IID	0.937
10	4	3	32	IID	0.919
10	4	4	32	IID	0.921
10	4	5	32	IID	0.924
10	1	3	16	IID	0.944

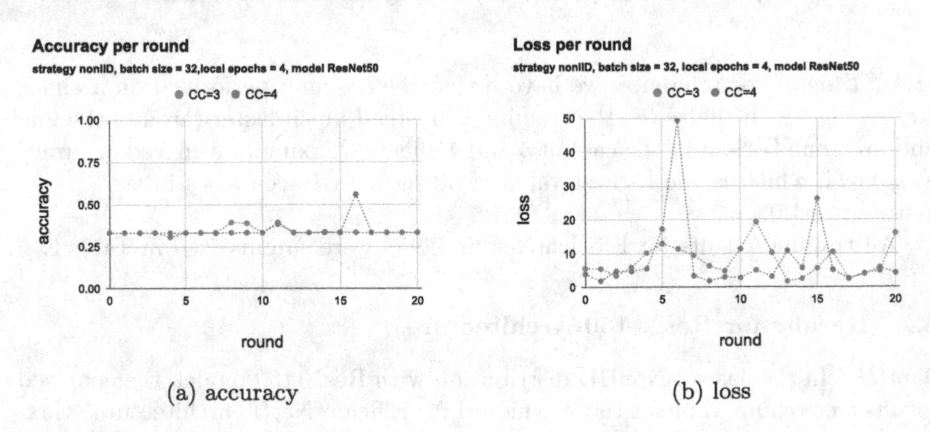

(a) accuracy (b) loss

Fig. 6. Model ResNet50, strategy NonIID, 3 and 4 clients

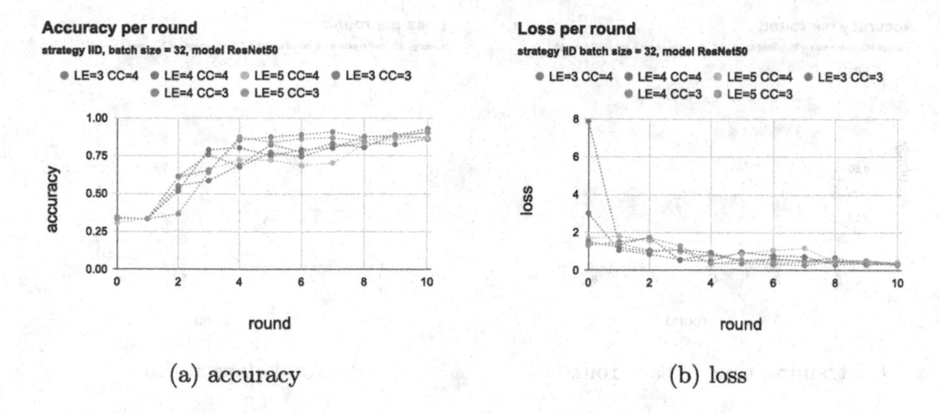

(a) accuracy (b) loss

Fig. 7. Model ResNet50, strategy IID, 3 and 4 clients

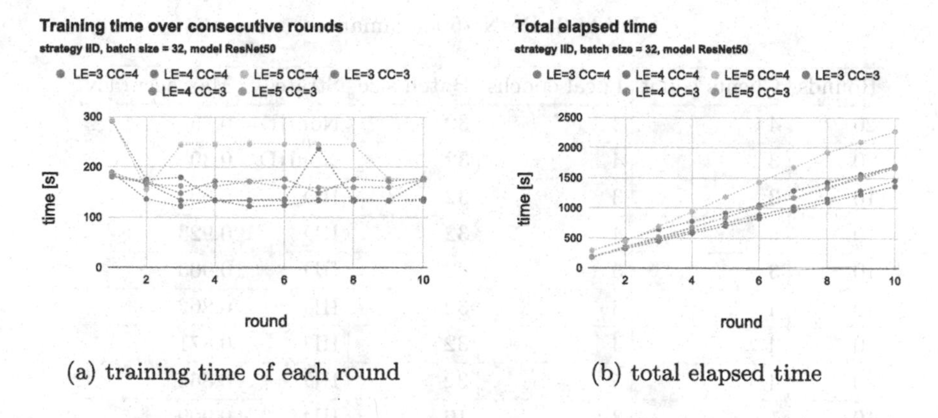

(a) training time of each round (b) total elapsed time

Fig. 8. Model ResNet50, strategy IID, 3 and 4 clients, training time chart

3 clients, increasing the number of local epochs resulted in faster convergence of training. In the case of 4 clients, 4 local epochs resulted in the best accuracy.

Additionally, training times were measured for 3 and 4 clients (Fig. 8). Increasing the number of local epochs resulted in a longer overall training time in the case of 4 clients. The fact that the training times for 4 local epochs and 5 local epochs in the case of 3 clients were similar was unanticipated. This could be attributed to a lower availability of GPU resources on Google Collab platform as the measurements were performed on different days and hours. Duration of the first round was almost always the longest, because of the construction of the execution graph in TensorFlow framework at the beginning of the training.

IID 1 Client: For reference, we have also tested the training performed on one client for ResNet50 model (presented in Fig. 9). Maximal accuracy was achieved only in the last training round. The process did not converge as smoothly as it was the case for the EfficientNetB0 model. The maximal accuracy was 0.909.

All training results for ResNet50 model were summarized in Table 2.

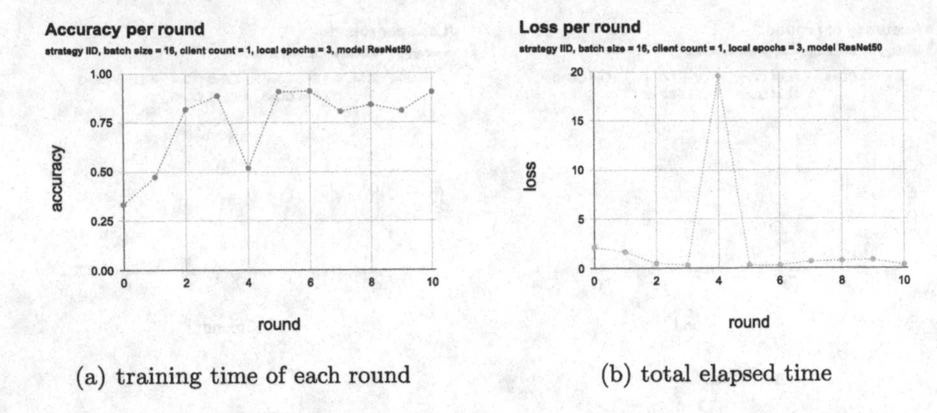

(a) training time of each round (b) total elapsed time

Fig. 9. Model ResNet50, strategy IID, 1 client

Table 2. ResNet50 summary

Rounds	Clients count	Local epochs	Batch size	Strategy	Max accuracy
20	4	4	32	NonIID	0.56
20	3	4	32	NonIID	0.39
10	3	3	32	IID	0.905
10	3	4	32	IID	0.923
10	3	5	32	IID	0.905
10	4	3	32	IID	0.862
10	4	4	32	IID	0.871
10	4	5	32	IID	0.868
10	1	3	16	IID	0.909

7 Conclusion

Despite the limited number of GPUs used, some conclusions can be drawn from the performed computations. The results on IID data outperformed models trained on nonIID datasets. What is more, EfficientNetB0 achieved a higher accuracy in comparison with ResNet50 regardless of the parameters. The only exception was training on nonIID datasets. Furthermore, as expected, the models trained on a single client scored higher accuracy on the test set. Another conclusion is that EfficientNetB0's score improved with the increase of the number of local epochs, whereas ResNet50 had the highest accuracy for 4 local epochs. Nonetheless, the more local epochs are set, the more GPU time is required for each round, however, the observed rise in the accuracy score is not significant.

To conclude, Federated Learning turns out to be a useful approach that allows for not only ensuring a proper level of security of computation but also training efficient models. Further research includes other areas where FL could be utilised, such as text classification or time series prediction.

Acknowledgments. This publication is partly supported by the EU H2020 grant "Sano" No 857533, and by the project "Sano" carried out within the International Research Agendas Programme of the Foundation for Polish Science, co-financed by the European Regional Development Fund.

References

1. Covid-19 radiography database. https://www.kaggle.com/tawsifurrahman/covid19-radiography-database
2. GRPC: A high performance, open source universal RPC framework. https://grpc.io/
3. Pygrid. https://github.com/OpenMined/PyGrid
4. PySyft. https://github.com/OpenMined/PySyft
5. Tensorflow federated. https://www.tensorflow.org/federated
6. Beutel, D.J., et al.: Flower: a friendly federated learning research framework (2021)
7. Dou, Q., et al.: Federated deep learning for detecting COVID-19 lung abnormalities in CT: a privacy-preserving multinational validation study. NPJ Digit. Med. **4**(1), 60 (2021). https://doi.org/10.1038/s41746-021-00431-6, http://www.nature.com/articles/s41746-021-00431-6
8. Kaissis, G.A., Makowski, M.R., Rückert, D., Braren, R.F.: Secure, privacy-preserving and federated machine learning in medical imaging. Nat. Mach. Intell. **2**(6), 305–311 (2020)
9. Liu, B., Yan, B., Zhou, Y., Yang, Y., Zhang, Y.: Experiments of federated learning for COVID-19 chest X-ray images (2020)
10. McMahan, H.B., Moore, E., Ramage, D., Hampson, S.: y Arcas. Communication-efficient learning of deep networks from decentralized data, B.A. (2017)
11. Ryffel, T., et al.: A generic framework for privacy preserving deep learning. CoRR abs/1811.04017 (2018). http://arxiv.org/abs/1811.04017
12. Sarma, K.V., et al.: Federated learning improves site performance in multicenter deep learning without data sharing. J. Am. Med. Inform. Assoc. **28**(6), 1259-1264 (2021). https://doi.org/10.1093/jamia/ocaa341, https://academic.oup.com/jamia/advance-article/doi/10.1093/jamia/ocaa341/6127556
13. Sheller, M.J., et al.: Federated learning in medicine: facilitating multi-institutional collaborations without sharing patient data. Sci. Rep. **10**(1), 1–12 (2020)
14. Silva, S., Altmann, A., Gutman, B., Lorenzi, M.: Fed-biomed: a general open-source frontend framework for federated learning in healthcare. In: Albarqouni, S., et al. (eds.) Domain Adaptation and Representation Transfer, and Distributed and Collaborative Learning, pp. 201–210. Springer International Publishing, Cham (2020). https://doi.org/10.1007/978-3-030-60548-3_20

On the Fairness of Swarm Learning in Skin Lesion Classification

Di Fan[1], Yifan Wu[2], and Xiaoxiao Li[3(✉)]

[1] University of California, Irvine, CA, USA
[2] University of Pennsylvania, Philadelphia, PA, USA
[3] The University of British Columbia, Vancouver, BC, Canada
xiaoxiao.li@ece.ubc.ca

Abstract. Fairness is essential for trustworthy Artificial Intelligence (AI) in healthcare. However, the existing AI model may be biased in its decision marking. The bias induced by data itself, such as collecting data in subgroups only, can be mitigated by including more diversified data. Distributed and collaborative learning is an approach to involve training models in massive, heterogeneous, and distributed data sources, also known as nodes. In this work, we target on examining the fairness issue in Swarm Learning (SL), a recent edge-computing based decentralized machine learning approach, which is designed for heterogeneous illnesses detection in precision medicine. SL has achieved high performance in clinical applications, but no attempt has been made to evaluate if SL can improve fairness. To address the problem, we present an empirical study by comparing the fairness among single (node) training, SL, centralized training. Specifically, we evaluate on large public available skin lesion dataset, which contains samples from various subgroups. The experiments demonstrate that SL does not exacerbate the fairness problem compared to centralized training and improves both performance and fairness compared to single training. However, there still exists biases in SL model and the implementation of SL is more complex than the alternative two strategies.

1 Introduction

The success of deep learning can be partially attributed to data-driven methodologies that automatically recognize patterns in large amounts of data. Machine learning can theoretically be performed locally if adequate data and computer infrastructure are available. However, data collection is costly and data sharing, such as cloud computing, is strictly regulated due to the privacy concerns in the healthcare domain.

Distributed and collaborative learning has gained popularity as a viable solution to federate disparate data and computational resources to provide a unifying analysis platform. Federated learning (FL) [15,23] is a trending type of learning scheme that avoids centralizing data in model training. Local data owners (also known as nodes or clients) can train the private model locally before sending

C. Oyarzun Laura et al. (Eds.): CLIP/DCL/LL-COVID/PPML 2021, LNCS 12969, pp. 120–129, 2021.
https://doi.org/10.1007/978-3-030-90874-4_12

the model weights or gradients to the central server via FL. The central server then aggregates the shared model parameters to create a new global model, which it then broadcasts to each local client. Recently, a new format of collaborative learning has been proposed, called Swarm Learning (SL) [21]. Unlike FL, SL replaces the central server to coordinate model updates and parameters communication with the Swarm network, which secures data privacy through blockchain technique [17] (See Fig. 1(c)). Additionally, in SL, transactions can only be carried out by the nodes having been pre-authorized and new nodes can be enrolled dynamically through blockchain smart contract. Model parameters are exchanged in the Swarm network by encryption and aggregated to update a new model at each round. We will illustrate more on the SL system in Sect. 3.

The statistical heterogeneity in collaborative learning and the local model multi-step training and then aggregation-based distributed learning paradigm leaves the deep learning model vulnerable to biases in the model, as the model may only attach importance to the dominant subgroup(s), such as, user groups with *sensitive (protected) attributes* (*i.e.* age or sex) that are over-represented in the deep learning model [4,7]. Such fairness problems have been recently discussed in the federated learning framework [12]. However, as SL is a fresh new concept proposed in recent days, no attempt has been made to evaluate the fairness or model bias issues in SL. Knowing the performance gap in different subgroups can guide researchers and practitioners to seek methods for better performance and fairness trade-off. This motivates us to conduct evaluations on the fairness in medical imaging tasks with different models. Specifically, we aim to compare model bias in centralized learning (pooling data to a data center), SL, and training model with a subgroup of data only (Fig. 1). Without additional bias mitigation methods, we explore whether the fairness behavior of SL leans to centralized learning or training on the subgroup. To this end, we conduct experiments on the skin lesion classification task, as skin lesions occur in diverse populations and have a severe class imbalance.

Our highlights are summarized in threefold:

1. We implement SL for a new medical application—skin lesion classification.
2. To the best of our knowledge, we are the first to investigate fairness on SL.
3. We observe that SL is robust to heterogeneous demographic-specific data distributions on our task, and it does not degrade the performance and fairness of the model compared to classical centralized training.

2 Related Works

2.1 Collaborative Learning and Their Application on Healthcare

Collaborative learning is a decentralized/distributed system that uses the principle of remote execution, which entails distributing duplication of a machine learning algorithm to the sites where the data is stored. These sites or devices are also called nodes. The system then runs the training method locally before

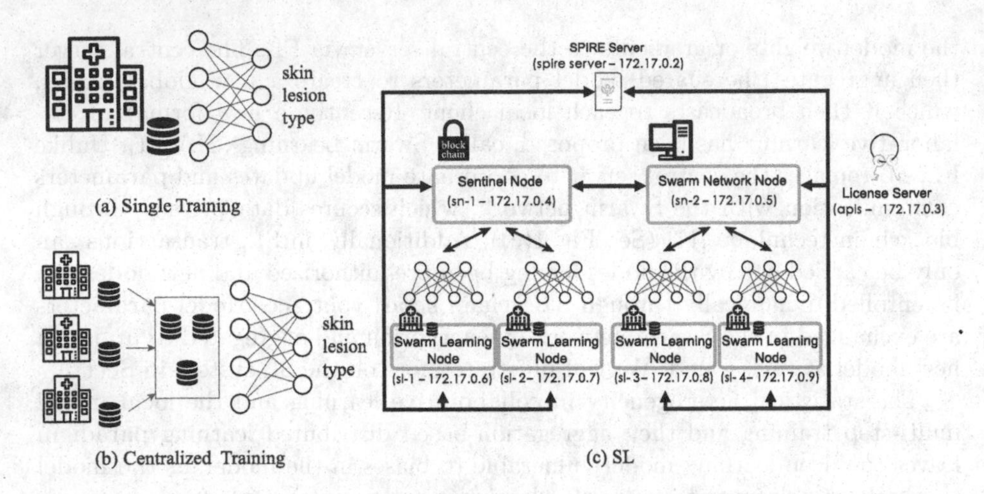

Fig. 1. Different training strategies to be evaluated in this study.

sending the results to a central repository to update the algorithm [9]. In the development of AI algorithms in the healthcare domain, data is one of the most important factors. However, there are two long-standing data-related challenges: 1) There are no standardized patient records; 2) Patient data is subject to stringent rules and protection standards. As a result, the concept of federated learning has recently gained a lot of attention. In the field of medical imaging, federated learning's applicability and benefits have been explored, such as on the problem of whole-brain segmentation and brain tumor segmentation [16], as well as finding disease-related biomarkers on fMRI [14].

2.2 Security and Privacy of Federated Learning

In federated learning, security and privacy are critical, for example, in the communication process between nodes. The existing methods include 1. Differential privacy [6], which is a method of preserving a dataset's global statistical distribution while eliminating individually identifiable data; 2. Homomorphic encryption, which is a type of encryption that allows unencrypted data to be processed as if it were unencrypted data; 3. Secure multi-party computation [24], meaning the process is done on encrypted data shares that are distributed among them in such a way that no single party can get all of the data on their own.

2.3 Fairness

Data inherently is biased. Deep learning algorithms have a tendency to mimic, if not increase, data bias. For example, existing chest X-ray classifiers were discovered to have differences in performance between subgroups defined by distinct phenotypes [18]. Using gender-balanced datasets, a recent study found statistically significant disparities in performance on medical imaging-based diagnosis [11]. Darker skinned patients may be underrepresented [10,13] in existing dermatology datasets [3,20], similar to forms of prejudice associated to face recognition [22]. The effects of decision-making that is (partially) based on the values of biased qualities can be irrevocable or even lethal, especially in medical applications.

3 Problem Setting and Methods

3.1 Problem Setting

In this paper, we focus on skin lesion classification. Skin cancer is a server public healthcare problem. In the United States, over 5 million newly diagnosed cases every year [8]. Early detection of skin lesions will decrease the death rate and reduce medical costs. Dermoscopy is one of the most widely used skin imaging techniques to distinguish the lesion spots on skin [2]. Our classification task is based on dermoscopic images.

Specifically, we cluster the dermoscopic images into subgroups based on patients' age and sex, which mimic the demographic-specific data distributions in different data silos. We consider data is distributed and assume each institution only contains a certain subgroup of dermoscopic images. We evaluate the model performance and fairness on unseen testing data set. To investigate model performance and bias in SL, we conduct experiments on skin lesion classification tasks in SL, and compare it to centralized training (pulling data together) and signal node training.

Next, we will introduce the different training strategies.

3.2 Swarm Learning

Swarm Learning (SL) is a decentralized, privacy-preserving Machine Learning framework. It utilizes a dedicated server to train the Machine Learning models with distributed data sources in a blockchain-safe strategy. As shown in Fig. 1(c), a set of local nodes process their local training data respectively without sharing with each other to obtain an ML model collaboratively. Then, via a Swam network rather than a centralized server, the ML parameters or weights are shared by each individual node, and finally, a merging model is formed.

At the beginning of SL, each local node should enroll or register with a blockchain Swarm smart contract. Then, this one-time process could enable each node to record information from the contract, like obtaining the model and performing local training as soon as it meets the synchronization condition. Next, the Swarm application programming interface (API) makes parameters exchanging among nodes and merges to update a new model before a new round of training. Finally, after the final round of parameter sharing and merging, the framework will check the stop criterion and halt if the criterion is reached otherwise back to local training.

With the blockchain techniques, SL has the following characteristics and advantages: (1) Saving large medical data locally; (2) No need to exchange original data, thereby reducing data traffic; (3) Providing high-level data security and protecting model from attacks; (4) No need for a secure central network; and (5) Allowing all members to merge parameters with equal rights.

3.3 Local and Centralized Training

Now, we describe two conventional training strategies to be compared with SL.

Single (Node) Training assumes no inter-institutional model and data sharing. The institution uses the local data to train deep learning models.

Centralized Training collects the distributed data in the central database or cloud server and trains deep learning models following the classical approach.

3.4 Fairness Definition and Metrics

In this section, we introduce the notation of fairness in classification model and the fairness evaluation metrics to be used in this study.

Definition 1 (Fair classifier). *Let denote binarized sensitive attribute* $z \in \{0, 1\}$ *that induces bias or unfairness. We define a classifier* f *with respect to data distribution* P *on* $\{(X, Z, Y) : \mathbb{R}^d \times \mathcal{Z} \times \mathcal{C}\}$, *where* $\mathcal{C} = \{1, 2, \ldots, C\}$ *and* C *is the number of classes. The classifier* f *is fair if*

$$P(f(x, z) \in \mathcal{C}|z = 0) = P(f(x, z) \in \mathcal{C}|z = 1)$$

Namely, if the distance between the two output distributions of $z = 0$ and $z = 1$ is zero, then f is claimed as a fair classifier.

Before introducing the fairness metrics, we first give the formula of true positive rate (TPR) and false positive rate (FPR). Given x_i, y_i, z_i as input, label, and bias indicator for the ith sample in the dataset $\mathcal{D} = \{X, Y, Z\}$. We denote the data samples in class $c \in \mathcal{C}$ as $\mathcal{D}_c$. Let $\hat{y}_i$ represent the predicted label of sample i.

The widely used bias quantification metrics are *Statistical Parity Difference (SPD), and Equal opportunity difference (EOD)* [1,5,7]:

SPD measures the difference in the probability of positive outcome between the privileged and under-privileged groups:

$$SPD = \frac{1}{C}\sum_{c}^{C}\left(P_{(x_i,y_i,z_i)\in\mathcal{D}_c}(\hat{y}_i = y_i|z_i = 0) - P_{(x_i,y_i,z_i)\in\mathcal{D}_c}(\hat{y}_i = y_i|z_i = 1)\right).$$

EOD measures the difference in TPR for the privileged and under-privileged groups. TPR for a given class $c \in \mathcal{C}$ and subgroup z are defined as $TPR_z^c = P_{(x_i,y_i,z_i)\in\mathcal{D}_c}(\hat{y}_i = y_i|z_i = z)$.

$$EOD = \frac{1}{C}\sum_{c}^{C}(TPR_{z=0}^c - TPR_{z=1}^c).$$

4 Experiment and Results

4.1 Dataset

Our data comes from the public skin lesion analysis dataset, Skin ISIC 2018 [3,20]. However, we will divide the data into distributed portions to mimic multi-institutional data collection. The whole ISIC 2018 dataset consists of 327 actinic keratosis (AKIEC), 514 basal cell carcinoma (BCC), 115 dermatofibroma (DF), 1113 melanoma (MEL), 6705 nevus (NV), 1099 pigmented benign keratosis (BKL), 142 vascular lesions (VASC) smaples and in total 10015 RGB images. We resize the images to 224 × 224. We consider sensitive attributes of ISIC dataset are *age* ($\geq$60 and <60), and *sex* (male and female).

To evaluate our algorithms, we randomly split the dataset into a training set (80%) and a test set (20%). As the data is unbalanced across lesion classes, we augment 15×, 10×, 5×, 50×, 0×, 40×, 5× for AKIEC, BCC, DF, MEL, NV, BKL, and VASC, respectively. We divide the training data into four subgroups – Male and age $\geq$60; Male and age <60; Female and age $\geq$60; Female and age <60. Hence we model four institutions, and each owns the data from a subgroup. After augmentation, the population proportion of the four subgroups are shown in Fig. 2(a). For testing dataset, to evaluate the model more efficiently, we make the data nearly satisfy the condition of age <60: age >60 = 1:1 and male: female = 1:1 without augmentation over the four subgroups are shown in Fig. 2(b). Finally, Fig. 3 presents the heterogeneous sample distributions of the seven lesion types in the four subgroups (here, the four institutions).

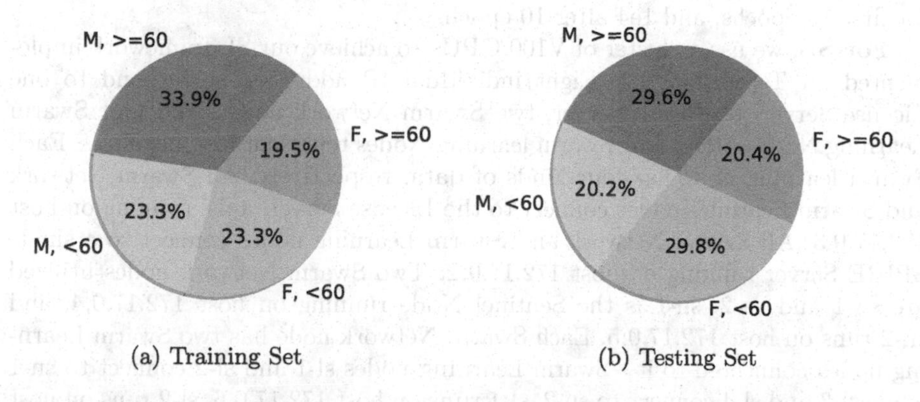

Fig. 2. Training data distribution (left), Test data distribution (right). The subgroup type is denoted next to the pie chart, where 'M' means 'male', 'F' means 'female', and ages are separated by <60 and $\geq$60.

4.2 Implementation Details

We use VGG19 [19] with cross-entropy loss to classify skin lesions among the 7 classes (AKIEC, BCC, DF, MEL, NV, BKL, and VASC). Batch size is set to

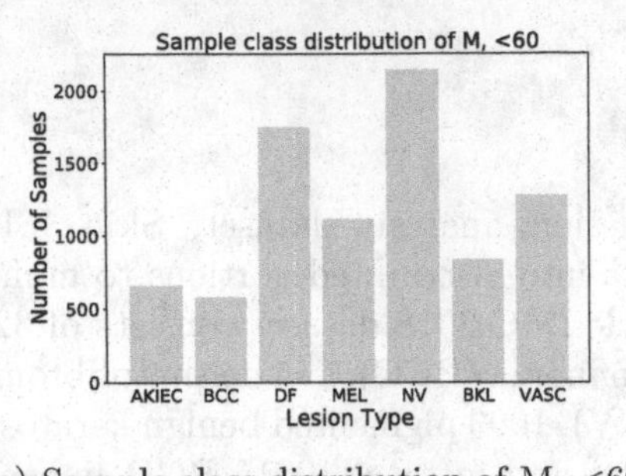

(a) Sample class distribution of M, <60

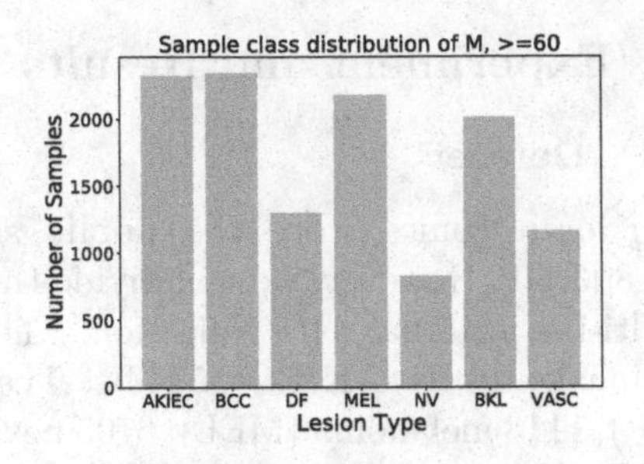

(b) Sample class distribution of M, >=60

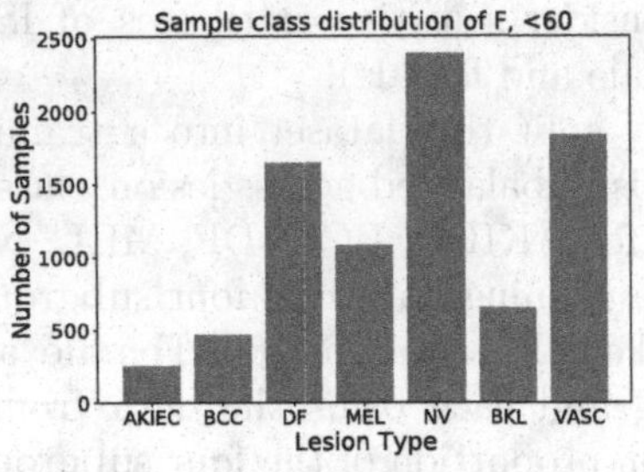

(c) Sample class distribution of F, <60

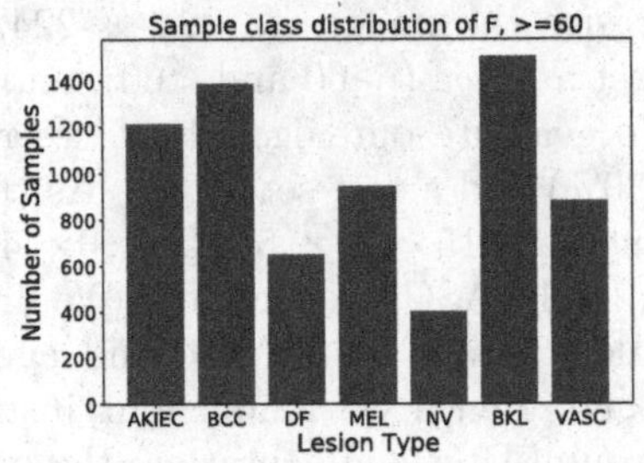

(d) Sample class distribution of F, >=60

Fig. 3. Label distributions in the four subgroups (institutions).

32. We use Adam optimizer, learning rate (lr) for classification training is 1e3 for first 10 epochs, and 1e4 after 10 epochs.

For SL, we use a cluster of V100 GPUs to achieve our SL framework implemented on Tensorflow 2.2. Eight individual IP addresses correspond to one License Server, one Spire Server, two Swarm Network nodes, and four Swarm Learning Nodes. Here, four Swarm learning Nodes represent four hospitals. Each Swarm learning node has four kinds of data, respectively. All Swarm Network and Swarm Learning nodes connect to the License Server, apls, running on host 172.17.0.3. All Swarm Network and Swarm Learning nodes connect to a single SPIRE Server running on host 172.17.0.2. Two Swarm Network nodes utilized are sn-1 and sn-2. sn-1 is the Sentinel Node running on host 172.17.0.4, and sn-2 runs on host 172.17.0.5. Each Swarm Network node has two Swarm Learning nodes connected to it—Swarm Learning nodes sl-1 and sl-2 connect to sn-1 while sl-3 and sl-4 connect to sn-2. sl-1 runs on host 172.17.0.6, sl-2 runs on host 172.17.0.7, sl-3 runs on host 172.17.0.8, sl-4 runs on host 1172.17.0.9.

For single training, model is trained on one V100 GPU using only one subgroup of data (M, age <60) representing a hospital.

For centralized training, we use the whole training dataset by pooling the four hospitals in SL together and train the data with one V100 GPU.

Table 1. Model performance. Higher scores indicate better performance.

	Sex	Precision	Recall	F1-score	Age	Precision	Recall	F1-score
Centralized	Male	0.774	0.680	0.707	≥60	0.812	0.752	0.776
	Female	0.899	0.893	0.892	<60	0.839	0.900	0.834
SL	Male	0.691	0.658	0.656	≥ 60	0.760	0.698	0.712
	Female	0.758	0.731	0.737	<60	0.638	0.762	0.660
Single	Male	0.579	0.460	0.498	≥60	0.578	0.428	0.471
	Female	0.786	0.547	0.617	<60	0.761	0.723	0.716

Table 2. Fairness scores. Lower SPD and EOD indicate less bias.

	Sex			Age		
	Centralized	SL	Single	Centralized	SL	Single
SPD	0.125	0.067	0.207	0.027	0.122	0.183
EOD	0.213	0.073	0.087	0.148	0.064	0.295

4.3 Biases in Models Trained with Different Strategies

We evaluate our framework on skin lesion detection against sensitive attribute: *age* and *sex*. The testing dataset contains samples from all the above four subgroups. We present the bias that existed in the centralized training, single training, and SL by showing the performance metrics (precision, recall, F1-score). The results of centralized training are listed in the *'Centralized'* row of Table 1. Row *'SL'* and row *'Single'* list the results for SL and single training respectively. The averaged precision, recall, F1-score are reported to evaluate the testing classification performance on the subgroups with different sensitive attributes. Centralized training achieved the best results. The result is within our expectation as the distributed optimization method in SL may hurt model performance, especially on heterogeneous data. Nevertheless, SL still outperforms single training and generalizes much better to testing data that contains unseen subgroups, on the average performances of both subgroups.

We notice the performance gaps between the subgroups exist for all the investigated training strategies, which reveals the biases. We denote the group with better classification performance (such as female, age <60) as privilege group $z = 0$ and the opposite group (such as male, age ≥60) as under-privileged group $z = 1$. We quantitatively measure the fairness scores (SPD, EOD) on the testing set for our models and the vanilla model, as shown in Table 2. Lower fairness scores indicate the model is less biased. We observe that SL achieves best results for fairness scores on 'sex' partition and EOD for 'age' partition, whereas the SPD for 'age' partition is less than centralized. In this case, SL achieves comparable performance and the same level of fairness in the heterogeneous demographic-specific data setting as centralized training. Also, centralized has a higher EOD for 'sex' partition than single and this could be in an explanation that single training has a poor performance for both 'male' and 'female' recall value in Table 1 and causes the difference to be small thus makes EOD lower compared to centralized.

5 Discussion and Conclusion

In this work, we make the first attempt to evaluate the model fairness in SL for medical applications. Specifically, we compare model performance and fairness with various metrics on single, centralized, and SL. Based on our experimental observation, SL could achieve better performance than training on a single institution, and the SL model does not amplify biases. However, we cannot conclude that SL gains the best performance-fairness trade-off for arbitrary clinical tasks. We also want to point out the high implementation complexity of SL framework due to the intricate blockchain network configurations. In summary, our work brings the interesting collaborative learning formula to skin cancer classification and designs experiments to evaluate its fairness medical applications and highlight. Our future work includes improving model performance for SL, investigating the mechanism of SL in handling model fairness, and designing bias mitigation methods for SL.

References

1. Bellamy, R.K., et al.: AI fairness 360: an extensible toolkit for detecting, understanding, and mitigating unwanted algorithmic bias. arXiv preprint arXiv:1810.01943 (2018)
2. Binder, M., et al.: Epiluminescence microscopy: a useful tool for the diagnosis of pigmented skin lesions for formally trained dermatologists. Arch. Dermatol. **131**(3), 286–291 (1995)
3. Codella, N.C., et al.: Skin lesion analysis toward melanoma detection: a challenge at the 2017 international symposium on biomedical imaging (ISBI), hosted by the international skin imaging collaboration (ISIC). In: 2018 IEEE 15th International Symposium on Biomedical Imaging (ISBI 2018), pp. 168–172. IEEE (2018)
4. Du, M., Yang, F., Zou, N., Hu, X.: Fairness in deep learning: a computational perspective. IEEE Intell. Syst. **36**, 25–34 (2020)
5. Dwork, C., Hardt, M., Pitassi, T., Reingold, O., Zemel, R.: Fairness through awareness. In: Proceedings of the 3rd Innovations in Theoretical Computer Science Conference, pp. 214–226 (2012)
6. Dwork, C., Roth, A., et al.: The algorithmic foundations of differential privacy. Found. Trends Theor. Comput. Sci. **9**(3–4), 211–407 (2014)
7. Hardt, M., Price, E., Srebro, N.: Equality of opportunity in supervised learning. Adv. Neural. Inf. Process. Syst. **29**, 3315–3323 (2016)
8. Jerant, A.F., Johnson, J.T., Sheridan, C.D., Caffrey, T.J.: Early detection and treatment of skin cancer. Am. Fam. Phys. **62**(2), 357–368 (2000)
9. Kaissis, G.A., Makowski, M.R., Rückert, D., Braren, R.F.: Secure, privacy-preserving and federated machine learning in medical imaging. Nat. Mach. Intell. **2**(6), 305–311 (2020)

10. Kinyanjui, N.M., et al.: Fairness of classifiers across skin tones in dermatology. In: Martel, A.L. (ed.) MICCAI 2020. LNCS, vol. 12266, pp. 320–329. Springer, Cham (2020). https://doi.org/10.1007/978-3-030-59725-2_31
11. Larrazabal, A.J., Nieto, N., Peterson, V., Milone, D.H., Ferrante, E.: Gender imbalance in medical imaging datasets produces biased classifiers for computer-aided diagnosis. Proc. Natl. Acad. Sci. **117**(23), 12592–12594 (2020)
12. Li, T., Sanjabi, M., Beirami, A., Smith, V.: Fair resource allocation in federated learning. arXiv preprint arXiv:1905.10497 (2019)
13. Li, X., Cui, Z., Wu, Y., Gu, L., Harada, T.: Estimating and improving fairness with adversarial learning. arXiv preprint arXiv:2103.04243 (2021)
14. Li, X., Gu, Y., Dvornek, N., Staib, L.H., Ventola, P., Duncan, J.S.: Multi-site fMRI analysis using privacy-preserving federated learning and domain adaptation: Abide results. Med. Image Anal. **65**, 101765 (2020)
15. McMahan, B., Moore, E., Ramage, D., Hampson, S., y Arcas, B.A.: Communication-efficient learning of deep networks from decentralized data. In: Proceedings of Artificial Intelligence and Statistics (AISTATS), pp. 1273–1282. PMLR (2017)
16. Rieke, N., et al.: The future of digital health with federated learning. NPJ Digit. Med. **3**(1), 1–7 (2020)
17. Saito, K., Yamada, H.: What's so different about blockchain?-blockchain is a probabilistic state machine. In: 2016 IEEE 36th International Conference on Distributed Computing Systems Workshops (ICDCSW), pp. 168–175. IEEE (2016)
18. Seyyed-Kalantari, L., Liu, G., McDermott, M., Chen, I.Y., Ghassemi, M.: CheXclusion: fairness gaps in deep chest X-ray classifiers (2020)
19. Simonyan, K., Zisserman, A.: Very deep convolutional networks for large-scale image recognition. arXiv preprint arXiv:1409.1556 (2014)
20. Tschandl, P., Rosendahl, C., Kittler, H.: The ham10000 dataset, a large collection of multi-source dermatoscopic images of common pigmented skin lesions. Sci. Data **5**, 180161 (2018)
21. Warnat-Herresthal, S., et al.: Swarm learning for decentralized and confidential clinical machine learning. Nature **594**(7862), 265–270 (2021)
22. Wu, Y., Yang, F., Xu, Y., Ling, H.: Privacy-protective-GAN for privacy preserving face de-identification. J. Comput. Sci. Technol. **34**, 47–60 (2019)
23. Yang, Q., Liu, Y., Chen, T., Tong, Y.: Federated machine learning: concept and applications. ACM Trans. Intell. Syst. Technol. (TIST) **10**(2), 12 (2019)
24. Zhao, C., et al.: Secure multi-party computation: theory, practice and applications. Inf. Sci. **476**, 357–372 (2019)

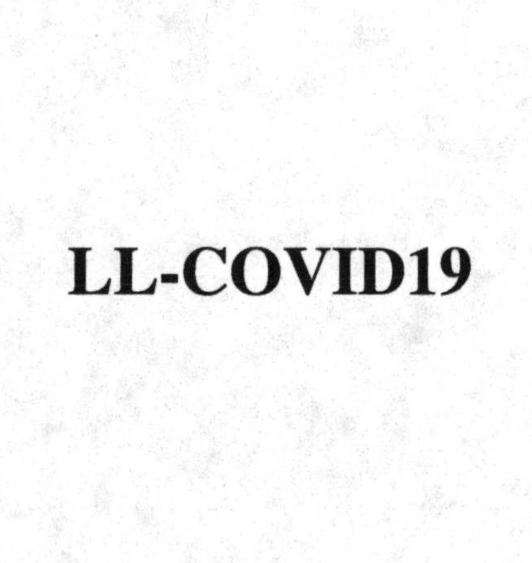

LL-COVID19

Lessons Learned from the Development and Application of Medical Imaging-Based AI Technologies for Combating COVID-19: Why Discuss, What Next

Maria Gabrani[1(✉)], Ender Konukoglu[2], David Beymer[3], Gustavo Carneiro[4], Jannis Born[1,5], Michal Guindy[6,7], and Michal Rosen-Zvi[8,9]

[1] IBM Research Europe, Zurich, Switzerland
mga@zurich.ibm.com
[2] Biomedical Image Computing, ETH, Zurich, Switzerland
[3] IBM Almaden Research Center, San Jose, CA, USA
[4] School of Computer Science at the University of Adelaide, Adelaide, Australia
[5] Department for Biosystems Science and Engineering, ETH, Zurich, Switzerland
[6] Assuta Medical Centres Radiology, Tel-Aviv, Israel
[7] Ben-Gurion University Medical School, Beersheba, Israel
[8] IBM Research Haifa, Haifa, Israel
rosen@il.ibm.com
[9] Faculty of Medicine, The Hebrew University of Jerusalem, Jerusalem, Israel

Abstract. The global COVID-19 pandemic has resulted in huge pressures on healthcare systems, with lung imaging, from chest radiographs (CXR) to computed tomography (CT) and ultrasound (US) of the thorax, playing an important role in the diagnosis and management of patients with coronavirus infection. The AI community reacted rapidly to the threat of the coronavirus pandemic by contributing numerous initiatives of developing AI technologies for interpreting lung images across the different modalities. We performed a thorough review of all relevant publications in 2020 [1] and identified numerous trends and insights that may help in accelerating the translation of AI technology in clinical practice in pandemic times. This workshop is devoted to the lessons learned from this accelerated process and in paving the way for further AI adoption.

In particular, the objective is to bring together radiologists and AI experts to review the scientific progress in the development of AI technologies for medical imaging to address the COVID-19 pandemic and share observations regarding the data relevance, the data availability and the translational aspects of AI research and development. We aim at understanding if and what needs to be done differently in developing technologies of AI for lung images of COVID-19 patients, given the pressure of an unprecedented pandemic - which processes are working, which should be further adapted, and which approaches should be abandoned.

C. Oyarzun Laura et al. (Eds.): CLIP/DCL/LL-COVID/PPML 2021, LNCS 12969, pp. 133–140, 2021.
https://doi.org/10.1007/978-3-030-90874-4_13

Keywords: COVID-19 · AI · Medical lung imaging · CT · CXR/XRay · Ultrasound

1 Introduction

In recent years, AI solutions have shown to be proficient in assisting radiologists and clinicians in detecting diseases, assessing severity, automatically localizing and quantifying disease features or providing an automated assessment of disease prognosis and treatment response. Consequently, AI for medical imaging (MI) received extraordinary attention in 2020, as, during the COVID-19 pandemic, lung imaging took a key role in managing COVID-19 patients as well as complementing biomolecular testing methods. The need to save time, cost and lives accelerated the leverage of AI in MI without fully demonstrating its ability in remedying the COVID-19 disease.

In a recent study [1] we performed a thorough review of all relevant publications in 2020 and identified a number of interesting trends and insightful messages. The pandemic significantly influenced the number of papers on AI in MI for COVID-19 in 2020 [1] but the quality of the manuscripts varies significantly. We evaluated the quality of publications based on aspects such as: (i) choice of AI model architecture, (ii) diversity in data sources, (iii) choice of evaluation metrics, (iv) experimental rigor and model generalization, and (v) reproducibility. Based on our methodical manual review, only 2.7% (namely 12) highly mature studies were identified. Our study revealed a number of limitations in the leveraged datasets including limited size, lack of diversity, and imbalance in disease conditions. In many situations, the datasets represented a population of patients with higher prevalence of COVID-19 at the time of imaging, which does not reflect true disease prevalence. Further, the models were deemed sensitive to motion artifacts, and other subtypes of lesions or comorbidities which caused data distribution shifts. Most studies also utilized datasets from limited geographical locations thereby restricting generalization performance of the models in other geographies.

Regardless of the availability and quality of relevant data, one of the main challenges revealed by our study was the actual relevance of the AI research focus with respect to the utility of the models developed. While most of the AI models developed were for the task of COVID-19 detection, based on available CT and or CXR datasets, the patient care teams found lung lesion segmentation and severity prediction more beneficial. AI models for these tasks, though, suffered from a lack of data and metadata availability. While a recent study points to the potential of convincing the medical community to adopt MI-based COVID-19 disease diagnosis, follow-up and prognosis [2], the issue remains the same. Data availability drives the AI model development. However, the process needs to be reversed, by mapping a medical need into a well-defined task with a measurable and applicable outcome, which shifts the point to the timely selection and provision of the relevant data, and, further, underlines the need for a closer collaboration between the AI and the domain communities.

To evaluate the continuous relevance of the 2020 analysis, we created similar statistics for the first 8 months of 2021, as depicted in Fig. 1 and 2. Interestingly, we observe similar trends with the previous year. While the number of papers in MI and AI remains similar to 2021, the COVID-19 publications on both MI and AI for MI have increased. Lung imaging is still the main focus, addressing the needs of the pandemic, and flattening the focus on breast imaging. The number of preprints is reducing, indicating the advancement of the peer-review process.

It is the aim of both the AI and the clinical communities to leverage data-driven models in patient care; however, the translation of AI research to clinical practice has not met the clinical needs with the speed and efficacy levels expected. Beyond task relevance, very few publications were further deemed mature enough with respect to validation depth and reproducibility. A key characteristic that underpins highly mature studies is an interdisciplinary and often multi-national collaboration of medical professionals and AI researchers.

While the messages are pointing to significant shifts in the way AI research and development is performed, research team sizes, number of involved countries and ratio of international collaborations are shrinking [3]. The MICCAI workshop LL-COVID19 is designed to facilitate the discussion between the AI community and medical experts, focusing on lessons learned from the months of the pandemic regarding if and how the development of mature and conducive AI technologies on imaging data can be accelerated. In this paper we review three key pillars in such collaborations: 1. Data definition, that is, which modality is of value in the case of COVID-19 and if a single modality is sufficient. 2. Data availability, that is the challenge of quickly assembling relevant data of a new disease 3. Making the research translational, that is, enable easy adoption of AI technologies in patient care.

Resolving the above three points could lead to more mature and conducive technologies and potentially assist clinicians and radiologists in addressing pressing clinical decision support needs during this or any future pandemic.

2 Data Definition

The AI literature analysis indicated that the vast majority of papers (72%) addressed diagnostic tasks as opposed to other tasks, such as severity assessment, disease features localization and quantification or prognosis. Revealing was the wider use of CXR data (50%) that was commonly utilized for the detection task. This was critically driven by annotated data availability. Analysis of the medical related literature, on the other hand, identified that MI played a more significant role in severity estimation and treatment monitoring. This also aligns with the ACR guidelines regarding imaging as an inconclusive test for COVID-19 detection due to uncertainties in accuracy and risk of cross-contamination [4]. While COVID-19 detection has been explored as a classification task even in CXR and US images, severity assessment and treatment monitoring concede towards the usage of CT devices, that have the capability to image increased occurrence of consolidation, linear opacities, crazy-paving patterns as well as bronchial wall thickening, differentiating the severity of COVID-19 patients.

Other drivers for defining data relevance include management tasks such as ventilation equipment, bed allocation, and device availability. While CT may provide higher resolution images, patients need to be moved, introducing virus transmission exposedness as well as health complications risks to already frail patients. In such cases, alternative technologies, such as US have been explored (e.g. [5]). However, acquiring US images requires personnel training and potential protocol changes.

While accuracy levels of AI models based on single modality varied, a few studies demonstrated that combination of imaging and clinical data can increase accuracy levels comparable to senior radiologists (e.g., [6]). An interesting study revealed similar trends when integrating US images and clinical data [7]. Another study leveraged manual diseases airspace segmentation from CT for volumetric quantification from CXR [8]. Data relevance is thus more than a single imaging modality definition, but rather a multimodal patient information interpretation assessment.

3 Data Availability

As our review revealed, a major factor for the maturity of a publication was contributed to the size and diversity of the leveraged data. The majority of mature publications utilized data obtained from multiple hospitals containing about 500 to 5000 patients' imaging data. Both size and diversity indeed require the collection and integration of data from multiple sources, which span lesion subtypes and disease conditions but also demographics, environmental and lifestyle practices, and point to international-level data availability and collections. A major step in this direction is open sourcing patient data. In the vast majority of the studies we analyzed, the data usually remained proprietary. However, at least partly, data were released in four mature papers [6,9–11].

With every institution having their own practices and protocols, harmonizing and standardizing data across different data sources may be a challenging task; however, methods such as transfer learning or domain adaptation may ease the task. A major issue though remains the one of data access, due to privacy or regulatory issues. While significant efforts are under way to both de-identify data and re-evaluate and adjust accordingly the regulatory requirements, AI development approaches such as federated learning may pave the way for integration of data and thus creation of models from larger and more diverse datasets.

Beyond images, metadata, such as labels or data from other modalities, such as clinical data, are of high value, to enable the training, testing and validation of the developed AI models. The datasets in our study were typically labelled using manual annotations from radiologists, RT-PCR tests and results from radiology reports. To cover larger, diverse datasets, crowd sourcing approaches are typically leveraged, but these are limited in the healthcare domain due to the knowledge and expertise required. There is certainly a large space for exploring such possibilities, through either training practices, or wider doctors' capacity utilization. Labelling by consensus or through an algorithmic approach, may further address the global issue of reducing bias and developing a globally accepted ground truth.

The quality of the data remains a major issue for any process, whether manual or automatic. But in demanding and strenuous times like in a pandemic, guaranteeing and leveraging high quality data, point to the need for both training of the professionals acquiring the data and to well established pipelines for data acquisition, management, storage and archiving, and thus to well designed, efficient digital workflows.

With respect to multimodal data, only three studies utilized clinical metadata in addition to images to develop their AI system [6,9,12]. With the alignment to clinical practice and medical recommendations [4], but also the demonstrated accuracy improvements [6,9,12,13] utilizing information from multiple modalities (e.g., imaging, clinical, molecular, pathology, etc.) identifies the need for multimodal data fusion. There are numerous initiatives that develop multimodal fusion approaches to integrate information from multiple modalities (e.g., [13]) that can further assist the task of data inclusion and coverage, however, with plenty of space for further research. One of the areas for further consideration in this direction, is the linkage and easy access of multimodal patient data, that typically span multiple departments and thus multiple silos. Digital workflows have the potential to address this aspect as well.

With timeliness of data availability being a major issue in building relevant AI models, all above enablers, from trained and dedicated data-acquisition and management personnel, to digital pipelines at both the hospitals and the AI technology institutions, and AI scale-out technologies, including federating learning, transfer and continuous learning, domain adaptation and multimodal data fusion, need to be re-evaluated and pushed forward.

4 Translational Research

Given the relevant datasets, and metadata, in abundant size and diversity, as well as the appropriate infrastructure, AI models can be developed as numerous publications have demonstrated (e.g., [1,2]). However, all these studies have also indicated that the translation of research to clinical practice has been moderate to low. Among all developed solutions, only two were deployed in clinical practice [14,15]. While many factors play a role, the special case of a pandemic has revealed shortcomings in experimental rigor, model interpretability and generalization and the need for closer collaboration between the AI and clinical communities. More specifically, we observed that while most publications reported confidence intervals and performed statistical tests, the evaluation processes were rather limited (e.g., on a single random split of the selected dataset). Most mature publications performed testing on external test datasets. Also, all of the mature publications used a human-in-the-loop (about 1 to 8 experienced radiologists) to compare and evaluate the proposed AI solutions, strengthening the value of their solution and building the trust between AI and AI users. Making such an evaluation scheme a standard practice, albeit time-consuming and demanding, paves the way for AI adoption. The same studies linked their model outputs to the input data (provided model explainability) by, for example, presenting heatmaps to illustrate regions of image the model build focused on. For clinical adoption, closing the gap between the AI-based recommendation and clinical domain expert knowledge and understanding is crucial.

Looking at another important translational dimension, namely model reuse, few studies conducted cross-validation and ablation studies to understand the generalization capabilities of their models. Only one solution was thoroughly tested in multiple countries [6]. On a positive note, a majority of the studies released the code publicly. Open sourcing is a major trend and an enabler to test numerous aspects of AI models and provide the necessary feedback, validation and robustness needed to build trust in AI models.

Finally, a key characteristic that underpins highly mature studies is an interdisciplinary and often multi-national collaboration of medical professionals and AI developers. Aligning the goals of diverse stakeholders (clinicians, AI experts, patients, funding and regulatory agencies etc.), and from the earliest stages of research, is one of the major cornerstones of translating research to clinical practice. Possible solutions include (i) inclusive execution and transparency (e.g., keep clinicians and/or patients involved throughout the process), (ii) robust evaluation of systems (e.g., going beyond accuracy metrics to incorporate reliability and usability metrics), and (iii) create common work environments.

Interestingly, beyond the ACR regulations that defined the procedures and management of COVID-19 patients [4] in medical practice, regulatory aspects, such as certification and clinical trials, were not the driving forces in the lack of AI adoption.

In summary, while in pandemics, time is crucial, special focus needs to be provided in understanding the impact of utilizing the models in practice. Rigorous testing and validation, model generalizability, explainability and open sourcing for further testing are key requirements. A one-team mindset between AI and domain experts, in co-designing the use case, the data availability, the model creation and validation is the key enabler.

5 Summary and Next Steps

The outbreak of the coronavirus pandemic presented a unique set-up for leveraging AI tools on lung modalities to the medical community. COVID-19 was a new condition, and clinicians and AI experts alike had to face this new condition and could form a joint understanding and design joint ways to address it. Particularly, as the number of cases has grown and confinement and many other non-pharmaceutical interventions were applied [16] and radiologists and other healthcare workers were using imaging differently complying with social-distancing rules, bedside modalities previously less commonly used have been preferred (e.g. [17]). Also the questions needing answers were changing fast. In the beginning the focus was on diagnosis of COVID-19, soon after, assessing severity has become a critical task. It seems that the long cycles typical to AI-based technology development have lead to a lag of the technology behind the clinical need. While, the sense of urgency the community have had, led to many novel collaborations between medical organizations and AI experts, it is likely that the lack of attentiveness by the physicians themselves who were busy taking

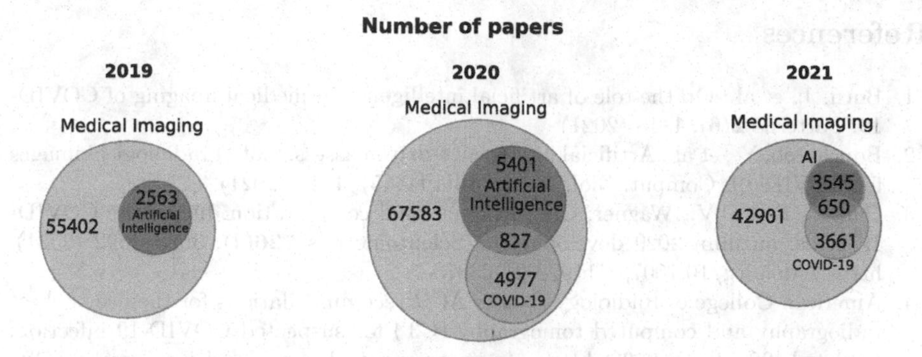

Fig. 1. Venn diagrams for AI in MI illustrating a publication increase between 2019 and 2021, at least partially due to the COVID-19 pandemic. Even in 2021, COVID-19 remains a key topic for the medical image analysis community. While the 2021 data is incomplete (reference date: 23.08.2021), the publication volume at the intersection of AI, COVID-19 and MI can be expected to keep growing in 2021, unlike the field of AI in MI in general. Diagrams created from automatic keyword searches on terms MI, AI, and COVID-19; on PubMed as well as preprint servers.

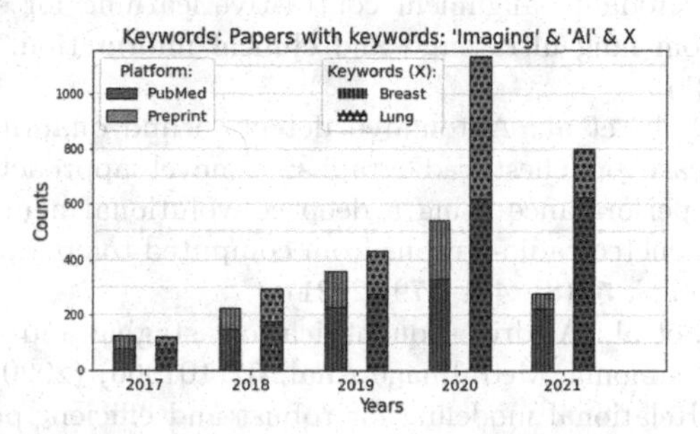

Fig. 2. Number of publications per keyword and platform. Paper counts using AI on breast or lung imaging since 2017. The bars for 2021 include all publications until 23.08.2021.

care of patients also played a role in the creation of the gap. On the other hand, the research and medical community's eagerness to share and publish every piece of news has probably led to the publication of a large number of sub-optimal often non-peer reviewed papers. Nowadays, as the knowledge of the condition is much wider than what was known in early 2020 and vaccine availability have enabled some control over the spread of the disease, it seems to be the right time for the community to have an open discussion of the lessons from the development and application of medical imaging-based AI technologies for combating COVID-19.

Acknowledgments. MRZ thanks the 8400 The Health Network and the Israeli Society for HealthTech for facilitating discussions with key opinion leaders on lessons learned.

References

1. Born, J., et al.: On the role of artificial intelligence in medical imaging of COVID-19. Patterns **2**(6), 1–18 (2021)
2. Bouchareb, Y., et al.: Artificial intelligence-driven assessment of radiological images for COVID-19. Comput. Biol. Med. **136**(104665), 1–17 (2021)
3. Cai, X., Fry, C.V., Wagner, C.S.: International collaboration during the COVID-19 crisis: autumn 2020 developments. Scientometrics **126**(4), 3683–3692 (2021). https://doi.org/10.1007/s11192-021-03873-7
4. American College of Radiology (2020): ACR recommendations for the use of chest radiography and computed tomography (CT) for suspected COVID-19 infection, Updated, 22 March 2020. https://www.acr.org/Advocacy-and-Economics/ACR-Position-Statements/Recommendations-for-Chest-Radiography-and-CT-for-Suspected-COVID19-Infection
5. Born, J., et al.: Accelerating detection of lung pathologies with explainable ultrasound image analysis. Appl. Sci. **11**(672) (2021)
6. Zhang, K., et al.: Clinically applicable AI system for accurate diagnosis, quantitative measurements, and prognosis of COVID-19 Pneumonia using computed tomography. Cell **181**(6), 1360 (2020)
7. Xue, W., et al.: Modality alignment contrastive learning for severity assessment of COVID-19 from lung ultrasound and clinical information. Med. Image Anal. **69**(101975) (2021)
8. Barbosa, E.J.M., Jr, et al.: Automated detection and quantification of COVID-19 airspace disease on chest radiographs: a novel approach achieving expert radiologist-level performance using a deep convolutional neural network trained on digital reconstructed radio- graphs from computed tomography-derived ground truth. Invest. Radiol. **56**(8), 471–479 (2021)
9. Chassagnon, G., et al.: AI-driven quantification, staging and outcome prediction of COVID-19 pneumonia. Med. Image Anal. **67**(101860) (2020)
10. Xie, W., et al.: Relational modeling for robust and efficient pulmonary lobe segmentation in CT scans. IEEE Trans. Med. Imaging **39**(8), 2664–2675 (2020)
11. Li, M.D., et al.: Automated assessment and tracking of COVID-19 pulmonary disease severity on chest radiographs using convolutional siamese neural networks. Radiol. Artif. Intell. **2**(4) (2020)
12. Mei, X., et al.: Artificial intelligence-enabled rapid diagnosis of patients with COVID-19. Nat. Med. **26**(8), 1224–1228 (2020)
13. Akselrod-Ballin, A., et al.: Predicting breast cancer by applying deep learning to linked health records and mammograms. Radiology **292**(2), 331–342 (2019)
14. Wang, B., et al.: AI-assisted CT imaging analysis for COVID-19 screening: building and deploying a medical AI system. Appl. Soft Comput. **98**(106897) (2020)
15. Gao, K., et al.: Dual-branch combination network (DCN): towards accurate diagnosis and lesion segmentation of COVID-19 using CT images. Med. Image Anal. **67**(101836) (2020)
16. Suryanarayanan, P., et al.: AI-assisted tracking of worldwide non-pharmaceutical interventions for COVID-19. Sci. Data **8**(1), 1–14 (2021)
17. Lu, W., et al.: A clinical study of noninvasive assessment of lung lesions in patients with coronavirus disease-19 (COVID-19) by bedside ultrasound. Ultraschall der Med.-Eur. J. Ultrasound **41**(03), 300–307 (2020)

The Role of Pleura and Adipose in Lung Ultrasound AI

Gautam Rajendrakumar Gare[1(✉)], Wanwen Chen[1], Alex Ling Yu Hung[1], Edward Chen[1], Hai V. Tran[2], Tom Fox[2], Peter Lowery[2], Kevin Zamora[2], Bennett P. DeBoisblanc[2], Ricardo Luis Rodriguez[3], and John Michael Galeotti[1]

[1] Robotics Institute and Department of ECE, Carnegie Mellon University, Pittsburgh, USA
ggare@andrew.cmu.edu

[2] Department of Pulmonary and Critical Care Medicine, Louisiana State University Health Sciences Center, New Orleans, USA

[3] Cosmeticsurg.net, LLC, Baltimore, USA

Abstract. In this paper, we study the significance of the pleura and adipose tissue in lung ultrasound AI analysis. We highlight their more prominent appearance when using high-frequency linear (HFL) instead of curvilinear ultrasound probes, showing HFL reveals better pleura detail. We compare the diagnostic utility of the pleura and adipose tissue using an HFL ultrasound probe. Masking the adipose tissue during training and inference (while retaining the pleural line and Merlin's space artifacts such as A-lines and B-lines) improved the AI model's diagnostic accuracy.

Keywords: Lung ultrasound · Pleura · Linear probe · Deep learning

1 Introduction

Point-of-care ultrasound (POCUS) is a non-invasive, real-time, bedside patient monitoring tool that is ideal for working with infectious diseases such as COVID-19. POCUS does not require the transport of critically ill contagious patients to a radiology suite, making it an easy choice for obtaining serial imaging to closely monitor disease progression. This has led to a significant interest in developing AI approaches for the interpretation of lung ultrasound (LUS) imaging [2,7,16,21].

A healthy lung is filled with air, which results in poor visibility of the internal anatomy in ultrasound images. Typical clinical practice for pulmonary ultrasound does not try to image the internal tissue of the lung but rather focuses on artifacts (e.g., "B-Lines") that are physically generated at the pleural membrane line. Traditionally, lung ultrasound in the intensive care unit (ICU) uses a low frequency (1–5 MHz) curvilinear or phased array (i.e., echocardiography) probe, which provides relatively deep and wide imaging. This approach is excellent for penetrating soft-tissues superficial to the lung and is standard practice

G. R. Gare and W. Chen—Equal contribution.

C. Oyarzun Laura et al. (Eds.): CLIP/DCL/LL-COVID/PPML 2021, LNCS 12969, pp. 141–149, 2021.
https://doi.org/10.1007/978-3-030-90874-4_14

for detecting prominent B-lines. Most AI research has followed typical clinical practice and focused on curvilinear probes and B-lines [16,21].

However, low-frequency curvilinear probes provide very poor detail of the pleural line. High-frequency-linear (HFL) ultrasound probes (typically in the 5–15 MHz range) offer a higher resolution of the pleural line but they can not image below 6–10 cm. Such reduced imaging depth is not a fundamental problem for lung ultrasound, except in cases of obesity or deep lung consolidation. In Fig. 1 we observe that on the curvilinear probe more depth is visible whereas on the linear probe the pleural line details are better seen. HFL probes are especially suitable for the easily-accessible L1 and R1 viewpoints [21], where the lung's visceral pleural membrane is more shallow.

Precise pleural line imaging may be more useful than deeper imaging penetration, even for pulmonary diseases that do not primarily manifest in the pleura. Carrer et al.'s automated method [4] to extract the pleural line and assess lung severity achieved better results from a linear probe compared to those from a curvilinear probe. Neonatal lung ultrasound analysis, which has also been extensively researched [10], typically calls for an HFL probe for its improved lung surface image quality. Although recent large datasets on COVID-19 [2,16,21] predominantly contain images from lower frequency curvilinear and phased array probes, they also include a small portion of linear probe images based on their observation that a linear probe is better for pleura visualization [2,18].

Since the HFL probe provides a better view of the pleural line, it can be a better option for COVID-19 lung ultrasound diagnosis. The SARS-CoV-2 virus that causes COVID-19 binds to Angiotensin-Converting Enzyme-2 (ACE-2) receptors of epithelial cells lining the bronchi and alveoli, and endothelial cells lining the pulmonary capillaries [14]. The lung injury from COVID-19 involves interalveolar septae that perpendicularly abut the visceral pleura. Therefore the pleural line should be a focus of the investigation. B-lines, which radiate deeply below the pleural line, have been extensively reported on. Although they are visualized below the pleural line, they are merely reverberation artifacts that emanate from within the pleural line. Thickening and disruption of the pleural line are subtle signs of underlying lung pathology that are poorly visualized using a curvilinear probe. In some cases these pleural line abnormalities are seen in the absence of B-lines and, therefore, such cases might be misclassified if a curvilinear probe were used. We propose that there are important and clinically relevant anatomic details visible in HFL images that are lacking in curvilinear images. Focusing on the pleura line itself, where pathology manifests the earliest, may yield clinically relevant information more directly than limiting interpretation to artifact appearance.

POCUS acoustic waves first propagate through the keratinized epidermis and the fibrous tissue and capillary networks of the dermis, then through adipose tissue and muscle bundles covered in fibrous fascia. The wavefront then traverses 1–4 microns of fibrous parietal pleura that lines the inside of the chest cavity before the lung's visceral pleura is reached. The acoustic characteristics of each of these structures are affected by probe location and patient characteristics, including age, sex, anatomy, lean body mass, and fat mass. Making use of the linear probe helps see not only the pleura but also the subcutaneous (SubQ) tissue structure in detail, which would otherwise occupy relatively few pixels with a curvilinear probe [13,20] (Refer to Fig. 1). This brings in additional challenges where a purported lung-AI network could instead rely on the SubQ to make the diagnosis rather than lung regions. AI might learn associations between these soft tissue structures and specific disease characteristics. For example, it has been well established that obesity and older age are risk factors for severe COVID-19 [6]. It would therefore be important that training and testing of AI approaches to the diagnosis of lung diseases consider the impact of the subcutaneous tissues.

We can broadly categorize the various regions that constitute the linear probe ultrasound image into the subcutaneous region, the pleura, and Merlin's space (i.e., real and artifact pixels beneath the pleural line) [11]. In the following sections of the paper, we try to determine the diagnostic prowess of these regions by generating images that emphasize these regions by masking out other regions. We study the diagnostic ability of the subcutaneous region (*subq*), subcutaneous+pleura (*subq+pleura*), the pleural region (*pleural*), the Merlin's region (*merlin*), and the pleural+Merlin's region (*pleural+merlin*). In addition, we also explore masking out indirect adipose/obesity information implicitly encoded by the depth and curvature of the pleura. So, we straighten the overall bend of the pleural line and mask out the depth by shifting up the pleura to a fixed distance from the top of the image. Refer to Fig. 3 for sample masked images.

2 Methodology

Problem Statement. Given an ultrasound B-mode scan clip I_g, the task is to find a function $F\colon [I_g] \rightarrow L$ that maps the clip I_g to ultrasound severity score labels $L \in \{0, 1, 2, 3\}$ as defined by [16]. Because the pleural line produces distinct artifacts (A-lines, B-lines) when scattering ultrasound based on the lung condition, the classification model should learn underlying mappings between the pleural line, artifacts, and pixel values, for making the predictions.

2.1 SubQ Masking

In lung ultrasound images, the SubQ tissue has more complicated tissue structures than those in the lung region. However, these structures might degrade the performance of AI-based diagnosis. The brighter and more complicated SubQ region has a larger response to the CNN layers than the lung region, but it does not provide much information on the underlying lung diseases and might even interfere with the performance of the deep neural network. So, to understand the role of the pleura and adipose in the COVID-19 diagnosis, we generate different masking of the LUS image regions (SubQ, pleural line, and Merlin's space).

Pleural Line Segmentation. The pleural line separates the SubQ tissue and Merlin's region in the ultrasound images and is usually the lower bright and wide horizontal line in the ultrasound images. To segment the pleural line, we first use a 5×5 Gaussian filter to blur the image and reduce the speckle noise. We then resize it to 150×150 to reduce the influence of the speckles in the segmentation. To find the candidate pixels that belong to a bright horizontal line, we first threshold the image based on the image's response to the Sobel filter along the y-axis with a 3×3 kernel size to find horizontal structures inside the images. We then threshold the image based on the intensity. We select the thresholds in Eq. 1 and Eq. 2, and we set $\alpha = 0.2$ and $\beta = 1.3$ in our dataset after tuning. In this way, we find the pixels that are on bright horizontal lines. Then in each column, we keep the lowest candidate point in the image, use dilation to fill the gap between the line. However, sometimes A-lines (which are reflections of the pleural line) can be mistakenly included in the candidate points. So we cluster the pixels into different regions based on the connectivity and keep the curve with the largest area as the main location of the pleural line. We then reflect potential A-lines back to the main location of the pleural line, to help fill in the pleural line as follows. We vertically shift other line segments to be at the same level with the pleural line, shifting each candidate pixel by an amount equal to the difference between the minimal y coordinate of the main location and the minimal y coordinates of the other curves. We then fit a fourth-order polynomial curve using the candidate pixels and extend the polynomial curve segment for more than 10 pixels along the tangent line at the two endpoints of the polynomial curve.

$$threshold_{sobel} = \alpha \times mean(I_{sobel}) \tag{1}$$

$$threshold_{intensity} = mean(I) + \beta \times std(I) \tag{2}$$

where I is the raw image, and I_{sobel} is the response of the image to Sobel filter.

With the segmented pleural line, the region above this line is the selected SubQ region, and the region below this line is the selected Merlin's region.

Pleural Line Straightening. We straighten and shift up the pleura in order to mask out the adipose/obesity information indirectly encoded into the curvature and depth of the pleura. Besides, different probe pressure would create different appearances of pleural lines in the images, so we want to eliminate the effect of this arbitrary variable as well. Therefore, we straighten the pleural lines while maintaining the local "bumps" on the pleural lines so that local pleura information would not be lost. In practice, we crop the images at 5 pixels above the pleural lines to preserve the information on the pleural lines and underneath them. We take the upper boundaries of the segmented pleural lines and fit a cubic function to it. (We did not use a higher-order function since we would like to preserve the local information on the pleural lines.) We then shift each column of the image upwards or downwards so that we make the cubic curve into a horizontal straight line. (Refer to Fig. 3 for sample straightened image.)

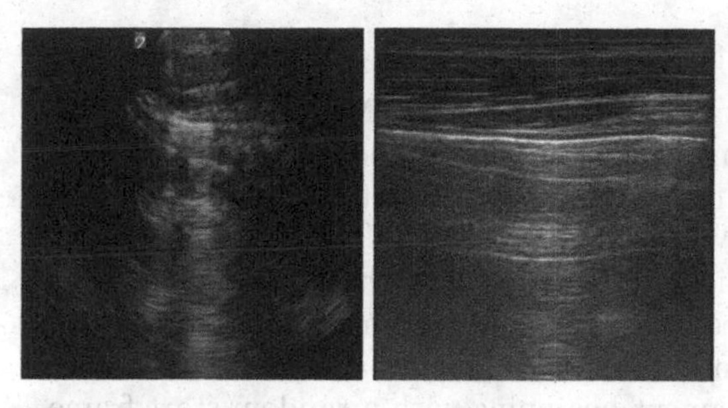

Fig. 1. Curvilinear (left) vs Linear (right) probe at the same **L1** position during a patient session. On the curvilinear, more depth is visible whereas, on the linear, pleural line details are better seen.

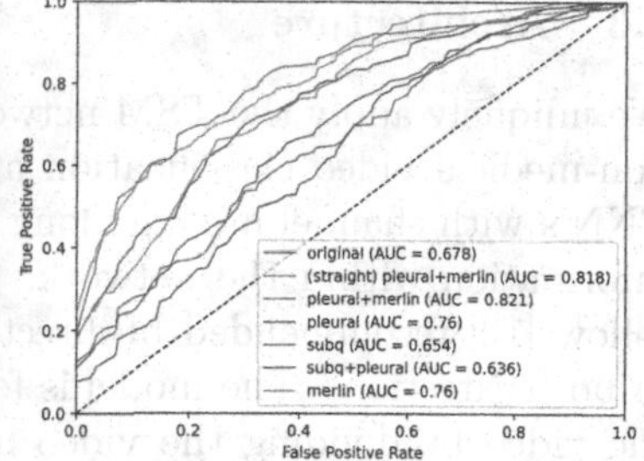

Fig. 2. RoC plots with AUC (macro averaged) of the models for video-based lung-severity scoring.

2.2 Data

Under IRB approval, we curated our own lung ultrasound dataset consisting of HFL probe videos. Our dataset consists of multiple ultrasound B-mode scans of L1 and R1 (left and right superior anterior) lung regions [21] at depths ranging from 4cm to 6cm under different scan settings, obtained using a Sonosite X-Porte ultrasound machine. The dataset consists of ultrasound scans of 93 unique patients from the pulmonary ED during COVID-19, and some patients were re-scanned on subsequent dates, yielding 210 videos.

We use the same 4-level ultrasound severity scoring scheme as defined in [1] which is similarly used in [16]. The score-0 indicates a normal lung with the presence of a continuous pleural line and horizontal A-line artifact. Scores 1 to 3 signify an abnormal lung, wherein score-1 indicates the presence of alterations in the pleural line with ≤ 5 vertical B-line artifacts, score-2 has the presence of >5 B-lines and score-3 signifies confounding B-lines with large consolidations (refer to [19] for sample images corresponding to the severity scores). All the manual labeling was performed by individuals with at least a month of training from a pulmonary ultrasound specialist. We have 27, 84, 75, and 24 videos labeled as scores 0, 1, 2, and 3 respectively.

Data Preprocessing. We perform dataset upsampling to address the class imbalance for the training data, wherein we upsample all the minority class labeled data to get a balanced training dataset [15]. All the images are resized to 224×224 pixels using bilinear interpolation. We augment the training data using random horizontal (left-to-right) flipping and scaling the image-pixel intensities by various scales $[0.8, 1.1]$.

2.3 Architecture

We uniquely apply the TSM network [12] to POCUS. TSM, commonly used for non-medical video classification and benchmarking methods, makes use of 2D CNN's with channel mixing along the temporal direction to infuse the temporal information within the network. We use ResNet-18 (RN18) [8] backbone and follow [12] recommended bi-directional residual shift with 1/8 channels shifted in both directions. The model is fed input clips of 18 frames wide sampled from the video by dividing the video into 18 equal segments and then selecting an equally spaced frame from each segment beginning with a random start frame.

2.4 Training Strategy

Implementation. The network is implemented with PyTorch and trained using the stochastic gradient descent algorithm [3] with an Adam optimizer [9] set with an initial learning rate of 0.001, to optimize over cross-entropy loss. The model is trained on an Nvidia Titan RTX GPU, with a batch size of 8 for 50 epochs. The ReduceLRonPlateau learning rate scheduler was used, which reduces the learning rate by a factor (0.5) when the performance metric (accuracy) plateaus on the validation set. For the final evaluation, we pick the best model with the highest validation set accuracy to test on the held-out test set.

Metrics. For the severity classification, we report accuracy and F1 score [2,16]. The receiver operating characteristic (ROC) curve is also reported along with its area under the curve (AUC) metric [2], wherein a weighted average is taken where the weights correspond to the support of each class and for the multi-label we consider the one-vs-all approach [5].

3 Experiments

We train the model on the various masked inputs and compare its performance to predict video-based lung-severity score labels. We randomly split the dataset into a training set and a separate held-out test set with 78%, and 22% split ratio respectively by randomly selecting videos while retaining the same distribution across the lung-severity scores in both the datasets and *ensuring no patient overlap between the test and train set.* Using the train set, we perform *5-fold cross-validation to create a training and validation fold.* We report the resulting metrics on the held-out test set in form of mean and standard deviation over the five independent cross-validation runs.

Table 1. Video-based 4-severity-level lung classification AUC of ROC, Accuracy, and F1 scores on a 93-patient HFL lung dataset. Highest scores shown in bold.

Method	AUC of ROC	Accuracy	F1-score
original	0.6553 ± 0.0425	0.4565 ± 0.0659	0.4381 ± 0.0701
subq	0.6154 ± 0.0619	0.4130 ± 0.0645	0.3656 ± 0.0956
pleural	0.7119 ± 0.0410	0.3783 ± 0.0525	0.3665 ± 0.0483
merlin	0.7076 ± 0.0436	0.4261 ± 0.0295	0.4183 ± 0.0299
subq+pleural	0.6303 ± 0.0560	0.3652 ± 0.0374	0.3204 ± 0.0381
pleural+merlin	**0.7742 ± 0.0648**	0.5261 ± 0.1178	0.5040 ± 0.1467
straightened pleural+merlin	0.7642 ± 0.0401	**0.5348 ± 0.0928**	**0.5166 ± 0.1016**

4 Results and Discussions

Table 1 shows the mean and standard deviation of the video-based severity scoring metrics, obtained by evaluating on the held-out test set using the models from the five independent runs. The two models with *pleural+merlin* input achieve the highest scores on all metrics, with the straightened version performing the best overall. The accuracy with the *pleura* is lower than the *subq* input, but combining the two gives the worst accuracy. The latter counter-intuitive result may be because the *subq* and *pleura* represent distinct diagnostic characteristics that AI may struggle to jointly model without seeing correlations in Merlin's space. Performance on the *original* image is inferior to the *pleural+merlin* image, perhaps because eliminating *subq* complexity makes it easier for the model to focus on the lung region to make diagnosis, as seen in Fig. 3. Individually, *merlin* has the best scores compared to *subq* and *pleura*. Combining *pleural+merlin* significantly improves the diagnostic accuracy of the model. The macro average RoC plots and AUC are shown in Fig. 2.

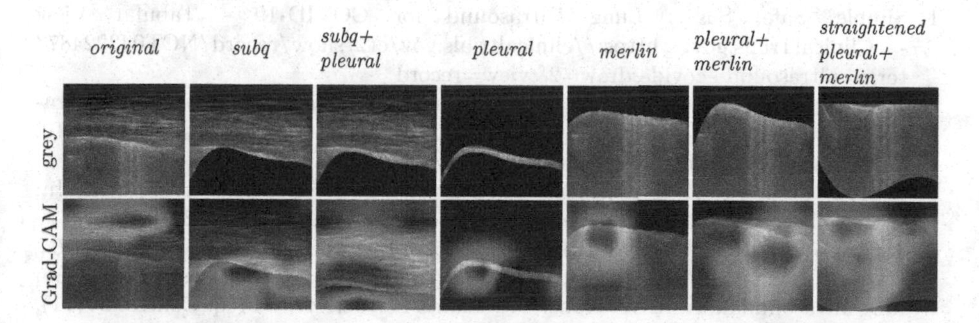

Fig. 3. Grad-CAM [17] visualization of layer 4 of the trained model on the various masked test images (B-mode grey). We observe that the model trained on *pleural+merlin* bases the predictions predominantly on the pleural line and B-line artifacts, whereas the *original* image trained model predominantly bases the predictions on the subcutaneous tissues above the pleural line.

Figure 3 depicts the various masked images of a frame from a test video. The Grad-CAM [17] visualization on the first video frame of the respective input trained model is also shown. This video was chosen because it represents trends that were qualitatively observed across the majority of videos. Both of the *pleural+merlin* models focused on Pleural line and B-line artifacts, whereas the *original*-image-input model focused on the SubQ region. The combination of *pleura+merlin* helped the model to focus on B-lines artifacts better than *merlin* alone. For this test video, all models except *subq* and *subq+pleura* correctly predicted the severity scores, suggesting that the *subq* is less diagnostically informative.

5 Conclusion

We highlighted the potential advantages of an HFL probe over the commonly used curvilinear probe in pulmonary ultrasound. We discussed the significance of having a well-imaged pleural line in addition to pleural artifacts, such as B-lines and A-lines, suggesting that carrying out AI analysis of the pleural line using linear probe could provide new avenues for carrying out challenging diagnoses. We demonstrated the diagnostic characteristics of the subcutaneous, pleura, and Merlin's-space regions of linear-probe ultrasound. From our experiments we draw that on masking out the subcutaneous region and retaining the detailed pleura along with Merlin's space has better diagnostic prowess.

Acknowledgements. This present work was sponsored in part by US Army Medical contract W81XWH-19-C0083. We are pursuing intellectual property protection. Galeotti serves on the advisory board of Activ Surgical, Inc. He and Rodriguez are involved in the startup Elio AI, Inc.

References

1. Simple, Safe, Same: Lung Ultrasound for COVID-19 - Tabular View - ClinicalTrials.gov. https://clinicaltrials.gov/ct2/show/record/NCT04322487?term=ultrasound+covid&draw=2&view=record
2. Born, J., et al.: Accelerating detection of lung pathologies with explainable ultrasound image analysis. Appl. Sci. (Switzerland) **11**(2) (2021). https://doi.org/10.3390/app11020672. https://www.mdpi.com/2076-3417/11/2/672
3. Bottou, L.: Large-scale machine learning with stochastic gradient descent. In: Lechevallier, Y., Saporta, G. (eds.) Proceedings of COMPSTAT'2010, pp. 177–186. Springer, Heidelberg (2010). https://doi.org/10.1007/978-3-7908-2604-3_16
4. Carrer, L., et al.: Automatic pleural line extraction and COVID-19 scoring from lung ultrasound data. IEEE Trans. Ultrasonics Ferroelectrics Freq. Control **67**(11), 2207–2217 (2020). https://doi.org/10.1109/TUFFC.2020.3005512
5. Fawcett, T.: An introduction to ROC analysis. Pattern Recognit. Lett. **27**(8), 861–874 (2006). https://doi.org/10.1016/j.patrec.2005.10.010
6. Gao, Y.D., et al.: Risk factors for severe and critically ill COVID-19 patients: a review. Allergy **76**(2), 428–455 (2021). https://doi.org/10.1111/ALL.14657. https://onlinelibrary.wiley.com/doi/full/10.1111/all.14657

7. Gare, G.R., et al.: Dense pixel-labeling for reverse-transfer and diagnostic learning on lung ultrasound for COVID-19 and pneumonia detection. In: Proceedings - International Symposium on Biomedical Imaging 2021-April, pp. 1406–1410 (2021). https://doi.org/10.1109/ISBI48211.2021.9433826
8. He, K., Zhang, X., Ren, S., Sun, J.: Deep residual learning for image recognition. In: Proceedings of the IEEE Computer Society Conference on Computer Vision and Pattern Recognition, vol. 2016-December, pp. 770–778. IEEE Computer Society (2016). https://doi.org/10.1109/CVPR.2016.90. http://image-net.org/challenges/LSVRC/2015/
9. Kingma, D.P., Ba, J.L.: Adam: a method for stochastic optimization. In: 3rd International Conference on Learning Representations, ICLR 2015 - Conference Track Proceedings. International Conference on Learning Representations, ICLR (2015). https://arxiv.org/abs/1412.6980v9
10. Liang, H.Y., et al.: Ultrasound in neonatal lung disease. Quan. Imaging Med, Surg. **8**(5), 535–546 (2018). https://doi.org/10.21037/qims.2018.06.01. https://www.ncbi.nlm.nih.gov/pmc/articles/PMC6037955/
11. Lichtenstein, D.: Novel approaches to ultrasonography of the lung and pleural space: where are we now? (2017). https://doi.org/10.1183/20734735.004717. https://pubmed.ncbi.nlm.nih.gov/28620429/
12. Lin, J., Gan, C., Han, S.: TSM: temporal shift module for efficient video understanding. In: Proceedings of the IEEE International Conference on Computer Vision 2019-Octob, pp. 7082–7092 (2018). http://arxiv.org/abs/1811.08383
13. Miller, A.: Practical approach to lung ultrasound. BJA Educ. **16**(2), 39–45 (2016). https://doi.org/10.1093/BJACEACCP/MKV012. https://academic.oup.com/bjaed/article/16/2/39/2897763
14. Ni, W., et al.: Role of angiotensin-converting enzyme 2 (ACE2) in COVID-19. Critical Care 2020 **24**(1), 1–10 (2020). https://doi.org/10.1186/S13054-020-03120-0. https://ccforum.biomedcentral.com/articles/10.1186/s13054-020-03120-0
15. Rahman, M.M., Davis, D.N.: Addressing the class imbalance problem in medical datasets. Int. J. Mach. Learn. Comput. 224–228 (2013). https://doi.org/10.7763/ijmlc.2013.v3.307
16. Roy, S., et al.: Deep learning for classification and localization of COVID-19 markers in point-of-care lung ultrasound. IEEE Trans. Med. Imaging **39**(8), 2676–2687 (2020). https://doi.org/10.1109/TMI.2020.2994459
17. Selvaraju, R.R., Cogswell, M., Das, A., Vedantam, R., Parikh, D., Batra, D.: Grad-CAM: visual explanations from deep networks via gradient-based localization. Int. J. Comput. Vis. **128**(2), 336–359 (2016). https://doi.org/10.1007/s11263-019-01228-7
18. Soldati, G., et al.: Is there a role for lung ultrasound during the COVID-19 pandemic? (2020). https://doi.org/10.1002/jum.15284. www.aium.org
19. Soldati, G., et al.: Proposal for international standardization of the use of lung ultrasound for patients with COVID-19. J. Ultrasound Med. **39**(7), 1413–1419 (2020). https://doi.org/10.1002/jum.15285. https://pubmed.ncbi.nlm.nih.gov/32227492/
20. Taylor, A., Anjum, F., O'Rourke, M.C.: Thoracic and lung ultrasound. StatPearls (2021). https://www.ncbi.nlm.nih.gov/books/NBK500013/
21. Xue, W., et al.: Modality alignment contrastive learning for severity assessment of COVID-19 from lung ultrasound and clinical information. Med. Image Anal. **69**, 101975 (2021). https://doi.org/10.1016/j.media.2021.101975

DuCN: Dual-Children Network for Medical Diagnosis and Similar Case Recommendation Towards COVID-19

Yunfei Long, Senhua Zhu(✉), and Dandan Tu(✉)

EI Innovation Lab, Huawei, Shanghai, China
{zhusenhua,tudandan}@huawei.com

Abstract. Early detection of the coronavirus disease 2019 (COVID-19) helps to treat patients timely and increase the cure rate, thus further suppressing the spread of the disease. In this study, we propose a novel deep learning based detection and similar case recommendation network to help control the epidemic. Our proposed network contains two stages: the first one is a lung region segmentation step and is used to exclude irrelevant factors, and the second is a detection and recommendation stage. Under this framework, in the second stage, we develop a dual-children network (DuCN) based on a pre-trained ResNet-18 to simultaneously realize the disease diagnosis and similar case recommendation. Besides, we employ triplet loss and intrapulmonary distance maps to assist the detection, which helps incorporate tiny differences between two images and is conducive to improving the diagnostic accuracy. For each confirmed COVID-19 case, we give similar cases to provide radiologists with diagnosis and treatment references. We conduct experiments on a large publicly available dataset (CC-CCII) and compare the proposed model with state-of-the-art COVID-19 detection methods. The results show that our proposed model achieves a promising clinical performance.

1 Introduction

Up to the present, the coronavirus disease 2019 (COVID-19) has caused massive infections and deaths over the world, and is still mutating rapidly. In recent years, deep learning (DL) based methods have made great achievements in lung disease analysis [4,17,18,24]. Also, many excellent works were proposed for COVID-19 detection [1,8,12,25] since 2019. The common DL-based COVID-19 detection methods usually employed a convolutional neural network (CNN) to extract image features and yielded predictions. For example, Jin et al. [13] designed a deep CNN detection system to diagnose COVID-19. Gao et al. [3] developed a dual-branch combination network, which combined the related lesion attention maps to assist the detection. Zhang et al. [28] proposed a diagnosis framework, which detected COVID-19 from other common pneumonias (CP) and normal healthy cases based on segmented lesion regions and achieved satisfactory results. Minaee et al. [19] used four pretrained models (ResNet18 [6], ResNet50 [6],

C. Oyarzun Laura et al. (Eds.): CLIP/DCL/LL-COVID/PPML 2021, LNCS 12969, pp. 150–159, 2021.
https://doi.org/10.1007/978-3-030-90874-4_15

SqueezeNet [11] and DenseNet-121 [10]) and deep transfer learning technique to do the detection. Ter-Sarkisov [23] introduced a lightweight Mask R-CNN [5] model to reduce the number of network parameters in COVID-19 detection. Hu et al. [9] developed a weakly supervised multi-scale learning framework, which assimilates different scales of lesion information for COVID-19 detection.

The above-mentioned methods were demonstrated to be effective and contributed a lot to combat COVID-19. However, the methods were proposed based on 2D images, which may omit important inter-slice information. To better extract representative features of COVID-19 lesions, several works [15,16,20,26, 29] proposed to diagnose the disease using 3D images. Among these approaches, Ouyang et al. [20] proposed to focus on the infection regions inside lungs, and then conducted the detection; while other works directly extracted features from entire images and make predictions. 3D methods usually outperforms 2D models since they incorporated more spatial features (e.g., inter-slice information) from 3D images. However, known 3D models were usually trained using the common cross-entropy loss, which may be strenuous to extract tiny differences between the novel coronavirus pneumonia (NCP) and CP . Moreover, none of the existing works provides similar cases for confirmed COVID-19 case, while similar cases may provide radiologists with significant treatment references.

In view of this, in this paper, we propose a dual-children network (DuCN) to simultaneously detect COVID-19 and provide similar cases for the confirmed case. In the proposed model, lung regions are segmented at first to exclude irrelevant regions hence eliminating the interference of irrelevant factors. Then, the segmented lung images are used for detection and recommendation. Meanwhile, the corresponding original full CT images and Euclidean distance maps in the lung regions are also integrated for providing abundant information. For COVID-19 detection process, we apply the triplet loss [22] to extract slight differences between NCP and CP. At last, once a case is confirmed with COVID-19, we bestow radiologists with similar cases, providing the diagnostic evidence and treatment references. Verified on a large clinical publicly available dataset (CC-CCII [28]) and compared with state-of-the-art methods, our new method yields promising results for COVID-19 detection and similar case recommendation.

Compared with previous COVID-19 detection works, our work makes the following contributions: (1) We develop a DuCN, which could give similar cases at the same time of diagnosis. To our best knowledge, this is the first work to provide COVID-19 similar cases; (2) We propose to use a triplet loss to supervise network to extract tiny differences between different type of images in COVID-19 detection; (3) We propose to use intrapulmonary Euclidean distance maps to assist incorporate more spatial information of infected lesions.

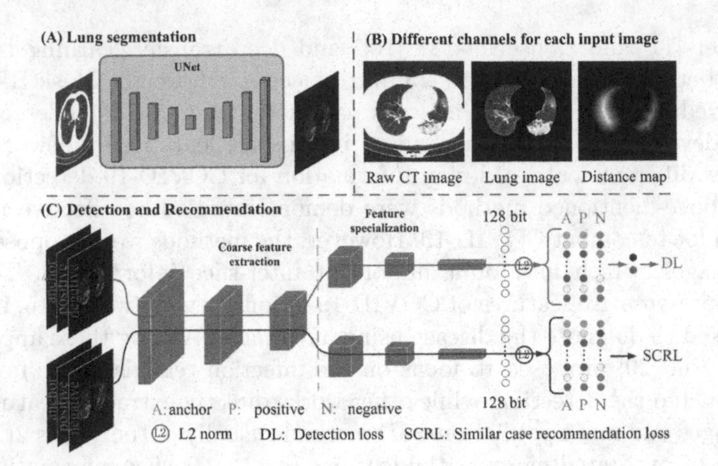

Fig. 1. Illustrating our proposed DuCN for medical diagnosis and similar case recommendation towards COVID-19.

2 Method

2.1 Proposed Model

Generally, the infection of the novel coronavirus mainly occurs in the lung area, and has little effect on areas outside lungs. Hence, we divide the detection and recommendation process into two stages: lung segmentation and detection/recommendation.

Figure 1 gives an overview of our proposed model. In the framework, firstly the original CT images are resized from 512×512 pixels to 224×224 pixels and input into a lung segmentation network, which is constructed by the common U-Net [21], and used to produce lung masks. The segmented lung masks are then combined with the corresponding original CT images to produce lung images. Besides, Euclidean distance maps in the lung regions are computed based on the segmented masks as in the previous work [17]. Finally, the dual-children network takes the original CT images, lung images and intrapulmonary Euclidean distance maps as inputs, and yields the probability of being infected with COVID-19. If a case is confirmed with COVID-19, the image-level similar cases are further provided. Below we present more details of our proposed model.

Someone may be conscious that our network has two more inputs (original CT images and intrapulmonary Euclidean distance maps) than previous works (e.g., [26]), which advocated using lung images to detect COVID-19. In general, if patients are infected with the novel coronavirus pneumonia, especially for critical patients, their liver will also be damaged [2]. Thus, the liver morphology in original CT images may promote the detection of the COVID-19. In this situation, using the full original CT images may be help make up for the information loss caused by probable inaccurate lung segmentation. For the extra input of intrapulmonary Euclidean distance maps, they are conducive to extracting more spatial information of infected lesions and improving the diagnosis/recommendation accuracy.

2.2 Dual-Children Network

The existing detection networks usually directly output the possibility of infection, without providing relevant diagnostic evidence or similar cases. However, in clinical practice, the reference of similar cases is of great significance to the treatment of diseases, especially for a new disease such as COVID-19. With this in mind, in this study, we develop a dual-children network (DuCN) to simultaneously detect COVID-19 and provide similar cases for confirmed patients.

Figure 1 (C) shows the proposed DuCN. As shown, the DuCN has two paths: the red path is designed for disease detection and the blue one is developed for similar case recommendation. For each path, we employ a pretrained ResNet18 [6] as backbone except that the output digits (1000) is replaced by 128. The two paths share network parameters in coarse feature extraction phase (the first three levels in ResNet18), and work independently in feature specialization phase (see Fig. 1 (C)). Please note that in the training and testing phases, the DuCN works differently. Below we elucidate the details for two different tasks respectively.

COVID-19 Detection. For COVID-19 detection, the previous works tended to use the common CrossEntropy loss to guide feature extraction. However, the CrossEntropy loss may be weak at picking up tiny differences between different types (some NCP and CP cases have very similar lesions). Triple loss [22], which was exploited for face recognition, could explore delicate changes between different cases due to the more comprehensive comparison between different type of cases. In this work, except for the CrossEntropy loss, to improve the detection sensitivity of the network, we employ the triplet loss to guide the delicate feature extraction. The triplet loss is defined as:

$$L_{triplet} = max\{d(f(a), f(p)) - d(f(a), f(n)) + margin, 0\} \quad (1)$$

where d represents the Euclidean distance; f is a feature extractor; a means an anchor sample, which is randomly selected from COVID-19 cases; p is an image (positive sample) of the same type as a, and is also randomly selected from COVID-19 cases, but from a different patient from a; n is a different type of negative sample from a, randomly selected from non-COVID-19 cases (including CP and normal cases in this study); $margin$ is the distance between positive and negative samples after training, we set it to 1.2 according to our numerical experiments.

The network for COVID-19 detection is presented in Fig. 1 (C) (the red path). In the training phase, the input for this task is triplet images (a, p and n), each of which contains three channels: lung image, full original CT image, and intrapulmonary Euclidean distance map. Each image in triplet images is processed by a ResNet18 (the red path in Fig. 1 (C)) in turn and outputs 128-bits one-dimensional features (3*128-bits in total). The extracted features are further normalized by an L_2 regularization and used for computing triplet loss. Then the features are transferred in 2-bits representations by a linear layer and used to compute CrossEntropy loss. After training, in the testing phase, the image to be detected (contains 3-channels) is feed in the network and generate two scores that represent the probability of different diseases.

Similar Case Recommendation. In clinical practice, doctors usually consult related similar confirmed cases to diagnose and treat existing cases, especially for some emerging or intractable diseases (e.g., COVID-19). Thus, devising an automated system to provide similar confirmed cases would be helpful for treating diseases. In this study, we develop a network to provide similar confirmed COVID-19 cases to help control the epidemic. The network is shown in Fig. 1 (C) (the blue path). In the training phase, the input is triplet images as in the detection path, the differences for the recommendation test are that p is randomly selected from the same patient as a, and n is randomly selected from a different patient but with the same disease (COVID-19) as a. This path is supervised by triplet loss only. After training, all images in the dataset are transferred into a 128-bit representations and saved as an index database. When an image is confirmed as COVID-19, it would be transferred into a 128-bit representations though the trained model and compared with samples in the library one by one using the Euclidean distance. The one with the smallest distance is the most similar case.

2.3 Loss Functions

Our proposed full network model has two sub-networks: U-Net and DuCN. The two sub-networks are trained separately, the segmentation network (U-Net) is trained using the common Dice loss, and DuCN is trained via the loss:

$$L_{total} = 0.3 * L_D + 0.7 * L_{SCR} \tag{2}$$

where $L_{SCR} = L_{triplet}$ is the similar case recommendation loss, and L_D is the COVID-19 detection loss ($L_D = 0.4 * L_{triplet} + 0.6 * L_{CrossEntropy}$).

3 Experiments and Results

3.1 Dataset and Experiments

For lung segmentation, we use a public dataset [14], which was collected for COVID-19 lung and lesion segmentation, to train the network. The dataset is split into 70% and 30% for training and testing, respectively. For COVID-19 detection and similar case recommendation, we employ a large dataset (CC-CCII [28]) to evaluate the proposed DuCN. We use data augmentation (flip left and right, and rotate ±2°) to expand the number of NCP images to a level equivalent to the number of non-NCP images. We split the dataset in patient-level (prevent CT images of the same patient from appearing simultaneously in training and testing sets) into 70% and 30% for training and testing, respectively.

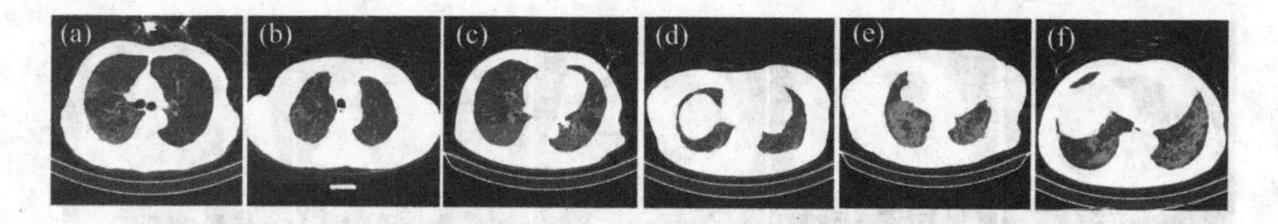

Fig. 2. Some representative visual segmentation results for lung regions.

Table 1. Quantitative results and comparison of different models. All the values are computed based on the entire testing dataset and are reported on a percentage scale.

Methods	Sensitivity	Specificity	Precision	F1-score	Accuracy	AUC
[28]	94.93	91.13	N/A	N/A	92.49	**97.97**
CovidNet-S [7]	91.72	N/A	88.78	90.23	88.55	N/A
CovidNet-L [7]	88.08	N/A	90.48	89.26	88.69	N/A
[27]	N/A	N/A	N/A	N/A	86.60	96.80
[13]	N/A	N/A	N/A	N/A	N/A	92.99
DuCN-LMR	96.32	91.28	95.46	95.89	95.21	93.96
DuCN-DMR	99.67	**94.43**	98.13	98.89	97.86	96.98
DuCN-RIR	98.63	94.25	97.82	98.22	97.90	96.88
DuCN-UP	99.68	94.40	98.07	98.86	98.01	96.92
Proposed	**99.85**	94.41	**98.54**	**98.90**	**98.28**	97.13

The quantitative evaluation for this study includes: dice score (for segmentation); sensitivity, specificity, precision, accuracy, F-1 score and AUC (for detection); and subjective judgments (for segmentation and similar case recommendation).

In our experiments, the two networks (U-Net and DuCN) are both implemented by PyTorch (V1.2.0) and trained separately on four NVIDIA K80 GPUs (12GB) with a batch size of 256. We adopt the Adam optimizer to train the networks with an initial learning rate of 1×10^{-3}, divided by 2 every 20 epochs.

3.2 Results

Lung Segmentation. Lung segmentation plays a critical role in the following step for COVID-19 detection and similar case recommendation. In the first stage, our segmentation network produces results with a dice score of 0.9669 on the entire testing dataset. However, since the dataset used for training segmentation network is different from the dataset in the second stage, a high dice score does not mean that the segmentation model will perform well in the second stage. Therefore, we apply the trained segmentation network in CC-CCII and evaluate the performance.

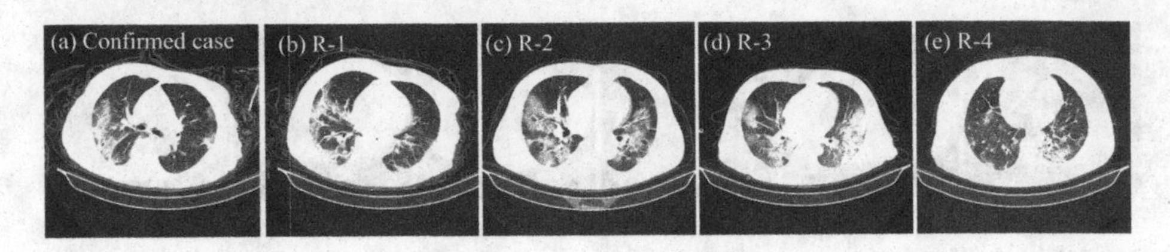

Fig. 3. Illustration of top-4 similar cases for confirmed COVID-19 case. R-N means the recommended case ranked N.

Figure 2 shows some representative visual segmentation results. Subjectively, the trained segmentation network performs very well, even for critical patients (e.g., see Figs. 2 (e-f)). The accurate segmentation of lung regions provides accurate and comprehensive lesion information for the subsequent step, and further promotes the accurate diagnosis of the disease.

COVID-19 Detection. In this work, we devise a DuCN to do COVID-19 detection and similar case recommendation. In this section, to verify the performance of the detection, we compare the results with five state-of-the-arts [7,13,27,28], among which [7] developed two models (CovidNet-S and CovidNet-L). The comparison methods were all conducted on the same publicly dataset: CC-CCII. Table 1 shows the values of various metrics for different methods, to ensure the fairness, we refer to the values reported in the original papers (N/A means that the original works did not use this indicator). From the comparison, it is clear that our new model produces promising results for COVID-19 detection.

Similar Case Recommendation. For confirmed cases, it is of great significance for treatment if doctors are provided relevant case references. Thus, our proposed DuCN gives similar cases at the same time of diagnosis. Since it is hard to evaluate such function quantitatively, we only assess it subjectively. Figure 3 presents a confirmed COVID-19 patient and its top-5 relevant cases. The top 1 is the patient itself, and for other three recommendations, the cases are related to the confirmed patient , implying that our similar case recommendation system is effective.

3.3 Ablation Study

Effectiveness of Raw Images, Lung masks and Distance Maps. For each slice of image that input into DuCN, it has 3-channels (a raw image, a lung image and a intrapulmonary distance map). To show the effectiveness of the 3-channels, we remove one of them from input. To ensure the input still contains 3 channels, if raw image or distance map removed (indicated as DuCN-RIR/DuCN-DMR), we replaced it with the corresponding lung image; if lung mask removed (indicated as DuCN-LMR), lung image and distance map are both replaced by the corresponding raw image. The results are illustrated in Table 1, from the results, all the three revisions has negative effect, suggesting that raw images, lung images and intrapulmonary Euclidean distance maps are all useful.

Does Pretrained Network Help? In this study, we use a pretrained ResNet-18 to construct DuCN. To test if the pretrained model help, we replace it in the full network with an untrained ResNet-18. The results is listed in Table 1 (DuCN-UP), one may observe that the pretrained network achieves a higher performance than the untrained one. Thus, it is reasonable to choose the pretrained ResNet18 to construct our DuCN.

4 Discussion and Conclusions

In this paper, we proposed a new DuCN to provide similar cases for confirmed COVID-19 patients at the same time of diagnosis. To exclude the interference of irrelevant factors outside lungs, we used a segmentation network to segment lung regions. Besides, the original CT images were incorporated to extract lesion related features (e.g., liver information). Further, intrepulmonary distance maps and triplet loss were introduced to help extract tiny differences. Validated on a large public dataset, our proposed network exhibits a promising performance in clinical application. Our proposed DuCN can be generically used in other disease screening applications. Still, our method has some shortcomings. For instance, for similar case recommendation, we selected positive samples from the same patient as anchor images for network training, which was not religious. In conclusion, a feasible method for the detection of COVID-19 is proposed.

References

1. Bai, H.X., et al.: Artificial intelligence augmentation of radiologist performance in distinguishing COVID-19 from pneumonia of other origin at chest CT. Radiology **296**(3), E156–E165 (2020)
2. Fan, Z., et al.: Clinical features of COVID-19-related liver functional abnormality. Clin. Gastroenterol. Hepatol. **18**(7), 1561–1566 (2020)
3. Gao, K., et al.: Dual-branch combination network (DCN): towards accurate diagnosis and lesion segmentation of COVID-19 using CT images. MIA **67**, 101836 (2021)
4. Guan, Q., Huang, Y., Zhong, Z., Zheng, Z., Zheng, L., Yang, Y.: Thorax disease classification with attention guided convolutional neural network. Pattern Recogn. Lett. **131**, 38–45 (2020)
5. He, K., Gkioxari, G., Dollár, P., Girshick, R.: Mask r-cnn. In: Proceedings of the IEEE ICCV, pp. 2961–2969 (2017)
6. He, K., Zhang, X., Ren, S., Sun, J.: Deep residual learning for image recognition. In: Proceedings of the IEEE Conference on CVPR, pp. 770–778 (2016)
7. He, X., et al.: Automated model design and benchmarking of 3d deep learning models for COVID-19 detection with chest ct scans. arXiv preprint arXiv:2101.05442 (2021)
8. Hemdan, E.E.D., Shouman, M.A., Karar, M.E.: Covidx-net: a framework of deep learning classifiers to diagnose COVID-19 in x-ray images. arXiv preprint arXiv:2003.11055 (2020)

9. Hu, S., et al.: Weakly supervised deep learning for COVID-19 infection detection and classification from CT images. IEEE Access **8**, 118869–118883 (2020)
10. Huang, G., Liu, Z., Van Der Maaten, L., Weinberger, K.Q.: Densely connected convolutional networks. In: Proceedings of the IEEE Conference on CVPR, pp. 4700–4708 (2017)
11. Iandola, F.N., Han, S., Moskewicz, M.W., Ashraf, K., Dally, W.J., Keutzer, K.: Squeezenet: alexnet-level accuracy with 50x fewer parameters and <0.5 mb model size. arXiv preprint arXiv:1602.07360 (2016)
12. Javaheri, T., et al.: Covidctnet: An open-source deep learning approach to identify COVID-19 using ct image. arXiv preprint arXiv:2005.03059 (2020)
13. Jin, C., et al.: Development and evaluation of an artificial intelligence system for COVID-19 diagnosis. Nat. Commun. **11**(1), 1–14 (2020)
14. Jun, M., et al.: COVID-19 CT lung and infection segmentation dataset (April 2020). https://doi.org/10.5281/zenodo.3757476
15. Li, L., et al.: Using artificial intelligence to detect COVID-19 and community-acquired pneumonia based on pulmonary CT: evaluation of the diagnostic accuracy. Radiology **296**(2), E65–E71 (2020)
16. Li, Y., et al.: Efficient and effective training of COVID-19 classification networks with self-supervised dual-track learning to rank. IEEE JBHI **24**(10), 2787–2797 (2020)
17. Liang, X., Peng, C., Qiu, B., Li, B.: Dense networks with relative location awareness for thorax disease identification. Med. Phys. **46**(5), 2064–2073 (2019)
18. Liu, X., Wang, K., Wang, K., Chen, T., Zhang, K., Wang, G., et al.: KISEG: a three-stage segmentation framework for multi-level acceleration of chest CT Scans from COVID-19 patients. In: Martel, A.L. (ed.) MICCAI 2020. LNCS, vol. 12264, pp. 25–34. Springer, Cham (2020). https://doi.org/10.1007/978-3-030-59719-1_3
19. Minaee, S., Kafieh, R., Sonka, M., Soufi, G.J.: Deep-COVID: predicting COVID-19 from chest x-ray images using deep transfer learning. MIA **65**, 101794 (2020)
20. Ouyang, X., et al.: Dual-sampling attention network for diagnosis of COVID-19 from community acquired pneumonia. IEEE TMI **39**(8), 2595–2605 (2020)
21. Ronneberger, O., Fischer, P., Brox, T.: U-Net: convolutional networks for biomedical image segmentation. In: Navab, N., Hornegger, J., Wells, W.M., Frangi, A.F. (eds.) MICCAI 2015. LNCS, vol. 9351, pp. 234–241. Springer, Cham (2015). https://doi.org/10.1007/978-3-319-24574-4_28
22. Schroff, F., Kalenichenko, D., Philbin, J.: Facenet: a unified embedding for face recognition and clustering. In: Proceedings of the IEEE CVPR, pp. 815–823 (2015)
23. Ter-Sarkisov, A.: Lightweight model for the prediction of COVID-19 through the detection and segmentation of lesions in chest CT scans. medRxiv (2020)
24. Wang, H., Jia, H., Lu, L., Xia, Y.: Thorax-net: an attention regularized deep neural network for classification of thoracic diseases on chest radiography. IEEE JBHI **24**(2), 475–485 (2019)
25. Wang, L., Lin, Z.Q., Wong, A.: COVID-net: a tailored deep convolutional neural network design for detection of COVID-19 cases from chest x-ray images. Sci. Rep. **10**(1), 1–12 (2020)
26. Wang, X., et al.: A weakly-supervised framework for COVID-19 classification and lesion localization from chest CT. IEEE TMI **39**(8), 2615–2625 (2020)
27. Wu, X., Chen, C., Zhong, M., Wang, J., Shi, J.: COVID-al: the diagnosis of COVID-19 with deep active learning. MIA **68**, 101913 (2021)

28. Zhang, X., et al.: Clinically applicable AI system for accurate diagnosis, quantitative measurements, and prognosis of COVID-19 pneumonia using computed tomography. Cell **181**(6), 1423–1433 (2020)
29. Zheng, C., et al.: Deep learning-based detection for COVID-19 from chest CT using weak label. MedRxiv (2020)

PPML

Data Imputation and Reconstruction of Distributed Parkinson's Disease Clinical Assessments: A Comparative Evaluation of Two Aggregation Algorithms

Jonatan Reyes[1(✉)], Yiming Xiao[1,2], and Marta Kersten-Oertel[1,2]

[1] Department of Computer Science and Software Engineering, Gina Cody School of Engineering and Computer Science, Concordia University, Montréal, QC, Canada
j_yes@encs.concordia.ca

[2] PERFORM Center, Concordia University, Montréal, QC, Canada

Abstract. Clinical assessments are an integral part of the care and management of Parkinson's disease, but full spectral assessments are difficult to obtain consistently, especially at follow-up visits. To better understand the etiology and pathogenesis of the disease and to offer accurate prognosis and tailored treatment plans, data-driven computational methods that rely on a large amount of quality clinical assessments have been proposed. However, major limitations, such as privacy and security issues have hindered the greater impact of these techniques. Motivated by the advantages of distributed and collaborative learning, we explore data imputation and reconstruction of clinical scores from the Parkinson Progression Marker Initiative (PPMI) in a multi-center distributed learning environment and we evaluate the reconstruction performance with two aggregation algorithms: Federated Averaging and Precision-weighted Federated Learning (A US provisional patent application has been filed for protecting at least one part of the innovation disclosed in this article). Our results suggest that while the first algorithm provides accurate reconstruction, the latter can better handle data heterogeneity between centers, reaching up to 19% lower reconstruction error.

Keywords: Parkinson's disease · Distributed learning · Imputation · Clinical assessments · Federated averaging · Precision-weighted federated learning

1 Introduction

Parkinson's disease (PD) is the second most frequent neurodegenerative disorder worldwide, associated with both motor and non-motor symptoms. Due to dopamine depletion from the disease, classic motor symptoms involve tremor,

C. Oyarzun Laura et al. (Eds.): CLIP/DCL/LL-COVID/PPML 2021, LNCS 12969, pp. 163–173, 2021.
https://doi.org/10.1007/978-3-030-90874-4_16

rigidity and bradykinesia, severely affecting the patient's daily functions. In addition, psychiatric issues including compulsive behaviors and depression [22], cognitive decline, and sleep disorders can also affect PD patients [9]. As the complexity of the disease has become more evident than previously assumed, further research is crucial to better understand the disorder.

1.1 Clinical Assessments and Challenges

To assess the degree of impairment and clinical progression of PD, clinical assessments are crucial in the care and management of PD, and are conducted in the forms of questionnaires, interactive tests, and objective measurements to gather information regarding both motor and non-motor symptoms. With the aim to search for effective biomarkers for accurate diagnosis and prognosis of PD, various data-driven approaches have been proposed to link imaging or bio-sample data with these clinical evaluations through statistical or machine/deep learning methods [12].

The Parkinson Progression Marker Initiative (PPMI) [11], sponsored by Michael J. Fox Foundation for Parkinson Research, is a comprehensive public multi-center database (www.ppmi-info.org/data), which includes longitudinal imaging, genetic, biosample and clinical assessment data of large PD cohorts. The up-to-date information on the study can be found at www.ppmi-info.org. Due to missing paper records and the nature of clinical protocols, some tests cannot be performed on the patient at the time of the visits, thus it is not uncommon for there to be missing clinical scores, especially in follow-up visits. This is a common issue when pooling data from multiple centres for disease-related studies, and exclusion of subjects with incomplete records has been adopted in some studies. As PD is highly heterogeneous in disease progression among individuals [8], exclusion of subjects may introduce sample bias in the related statistical and machine/deep learning methods, resulting in unreliable insights. Often data imputation is a solution to the problem. In this technique missing values are replaced with estimations based on the interpretation of contextual information and population distribution [7]. Accurate data imputation of the clinical scores in PD will effectively ensure the reliability and accuracy of the related studies and the proposed machine learning algorithms.

Another major challenge in the application of clinical data-driven algorithms is associated to restrictions on sharing patient data between medical and research centers. Patient data is sensitive and cannot be disclosed as serious privacy issues may arise. For example, patient may be discriminated by employers, insurance companies, or peers based on their health condition, causing embarrassment, paranoia or mental pain [16]. Owing to this, the access to clinical data is limited to the amount of information available within the medical or research center. This may not be adequate for training optimal machine/deep learning algorithms for clinical tasks and biomarker discoveries as it often leads to biased analysis. This is especially relevant for rare diseases, where very few patients are seen at any single institution [5,20]. To alleviate these issues, emerging data aggregation techniques have been developed to combine data from multiple sources with special attention to security and privacy.

1.2 Contributions

In this paper, we compare two types of aggregation algorithms based on distributed and collaborative learning: Federated Averaging and Precision-weighted Federated Learning. We explore the task of data imputation and reconstruction of clinical scores for PD using the PPMI database and investigate the effects of: (1) the performance of distributed machine learning aggregation algorithms, and (2) the imputation and reconstruction performance varying the number of missing values in the training dataset. To the best of our knowledge, we are the first to evaluate the performance of Federated Averaging and Precision-weighted Federated Learning aggregation algorithms in distributed learning environments for the task of data imputation and reconstruction of PD clinical assessments.

2 Related Work

Previous studies have addressed the problem of data imputation in biomedical data from different angles. The simplest case is deletion, where an entire patient record is removed from the database in the presence of missing values. However, it has been demonstrated that this method degrades the statistical power and yields bias estimates [1]. A more sophisticated statistical method for handling missing values is the single imputation technique. An example of this method is the last observation carried forward (LOCF) approach, where missing values in a longitudinal study are replaced by the last observation recorded [10].

Other methods require the analysis of multiple instances of the data, such as is the case of multiple imputation (MI). MI creates multiple copies of the plausible imputed data sets and estimates associations between the aggregated results [19]. Similarly, the multivariate imputation by chained equations (MICE) creates multiple imputation predictions for each missing value [2]. This method was particularly used to prepare the PPMI data used by Long-Short Term Memory networks to define PD subtypes and predict symptom progression [23]. Alternatively, machine learning-based techniques have been used to compressed clinical assessments and estimate missing scores. Peralta *et al.* investigated data imputation and reconstruction performance of clinical assessments with autoencoders [15]. With FCAEs, the encoder and decoder layers are organized in a fully-connected fashion, which rely on residual substructures called "Computational Blocks".

3 Methods

3.1 Data

We included 17 primary clinical assessments (and their sub-scores) and factors of the PPMI dataset (Table 1). These clinical assessments evaluate both motor and non-motor symptoms of the disease, including motor dysfunctions, psychiatric issues, cognitive functioning. Compared to the work of Peralta *et al.* [15], we utilize half the number of features for the training of machine learning models.

To prepare for the data ingestion process, we performed feature scaling by applying min-max normalization to transform the input data into the range [0,1]. Categorical data were mapped to ordinal values. Crucial information to the diagnosis and prognosis of the disease, such as demographic data (e.g., age, sex, and education) and patients' genotypes were collected at baseline visits and considered constant responses across visits (with the exception of age that changed according to the year of the next assessment). Similar to [15], we excluded the level of dopamine (LEDD) and SPECT imaging data, and focused our work in the primary clinical assessments. The final database contained baseline clinical assessments of 678 subjects and their follow-up visits over 3 years, containing 2466 rows and 102 columns. Table 1 demonstrates the percentages of missing values per assessment.

Table 1. Clinical assessments taken from PPMI and the percentage of rows with missing scores

Questionnaire	% missing value	Questionnaire	% missing value
Benton JLO Test	0.64	SCOPA	1.09
Epworth Sleepiness	0.32	Semantic Fluency	0.40
Geriatric Depression Scale	0.24	Schwab & England ADL	28.66
Hopkins VLT	0.68	STAI	0.36
LNS	0.40	UPDRS I	0.36
MoCA	0.77	UPDRS IP	0.32
QUIP	0.28	UPDRS II	41.20
REM Behavior Disorder	0.40	UPDRS III	41.24
Symbol Digit Modalities Score	0.48		

Additionally, to simulate the training of multi-institutional models in a distributed learning environment, we split patients randomly into four cohorts. With each cohort assigned to a site, patient information remains independent and is never shared between centers. Given this setup, we reserve one of these sites as the test set to measure the model's generalization performance with patient data never used during training based on the accuracy of reconstruction estimations for missing and non-missing clinical scores.

3.2 Model Setup

Fully-Connected Autoencoders (FCAEs). We use FCAEs introduced by Peralta *et al.* [15]. The architecture is available in a public repository[1]. Each

[1] https://github.com/m-prl/PatiNAE.

FCAE is trained with a NAdam optimizer using an initial learning rate of 0.001. The NAdam algorithm is a variation of the Adam optimizer that implements Nesterov momentum that accelerates convergence. To control the amount of unnecessary computation per client, we apply a strategy to reduce the learning rate when the model reaches a plateau in learning and stop training when this plateau persists for more than 60 training passes. We use a Mean Squared Error (MSE) to minimize the loss function over the reconstruction estimations. FCAEs models are implemented with Keras 2.4.3 and Tensorflow 2.4.1.

For evaluation purposes, reconstruction estimations are evaluated on the test site based on the two accuracy measurements defined in [15]:

$$A_1 = \frac{1}{K}\sum_{i=1}^{N}\sum_{j=1}^{M}(\hat{x}_j^i - x_j^i)^2 * M_j^i \tag{1}$$

$$A_2 = \frac{1}{U}\sum_{i=1}^{N}\sum_{j=1}^{M}(\hat{x}_j^i - x_j^i)^2 * (1 - M_j^i) \tag{2}$$

where Eq. 1 measures the reconstruction performance for non-missing clinical scores based on the total number of known scores (K) in the database, and a mask that identifies non-missing values M_j^i, and where x_j^i represents individual clinical scores and their respective reconstruction estimation $\hat{x}_j^i$. Similarly, Eq. 2 quantifies the reconstruction performance of missing clinical scores given the total number of unknown clinical scores (U) and a mask that identifies missing values.

3.3 Aggregation Algorithms

Federated Averaging (FedAvg). Federated learning introduced by McMahan *et al.* [13] works in rounds of communication through a distributed batch of local devices to learn a shared global model. At the beginning of each round, a server sends the initial shared global model to every client. Then, every client uses the shared model to compute stochastic gradient descent (SGD) optimizations with the local data and the resulting update (e.g., network weights) is sent to the server for further processing. After receiving all individual local updates, the central server aggregates them via the FedAvg algorithm to update the parameters of the shared global model, such that:

$$w_{t+1} \leftarrow \sum_{k=1}^{K}\frac{n_k}{n}w_{t+1}^k. \tag{3}$$

where $w_t^k + 1$ denotes the model weights of client k at iteration t, n_k is the number of local training samples n_k and n is the total number of samples. The round of communication repeats and as more rounds of communications are performed with this setting, the model learns a better representation of the data distribution and thus performance of the shared global model is optimized. Furthermore, when a new client joins the round of communication, the global model contains enough information from other clients that there is no need to retrain the model as it can be used immediately on the new device.

Precision-Weighted Federated Learning (PW). In [18], we proposed the PW algorithm as a variance-based aggregation scheme that averages the weights of distributed machine learning models. This algorithm differs from FedAvg in the way that individual local updates are aggregated. Instead of using the ratio of data samples as the multiplicative factor for weight update, PW takes into account local variance estimations, which are computed by the optimizer, and the update of the parameters of the shared global model is made in proportion to the inverse of this variance:

$$w_{t+1} \leftarrow \sum_{k=1}^{K} \frac{\left(v_{t+1}^{k}\right)^{-1}}{\sum_{k=1}^{K}\left(v_{t+1}^{k}\right)^{-1}} w_{t+1}^{k} \tag{4}$$

where v_{t+1}^{k} denotes the estimated variance of a given weight w at iteration t for client k. This method has shown significant advantages when the data is highly-heterogeneous across clients for benchmark datasets (e.g., MNIST, Fashion-MNIST, and CIFAR-10). However, to the best of our knowledge, no prior studies have examined the performance of this algorithm with medical data.

4 Experimental Results

4.1 Effect of Number of Missing Modalities During Training

Given a *corruption ratio*, FCAEs implement a masking layer to remove an entire modality (i.e., a clinical test) from patient records at random. We vary the corruption ratio of the training data to show how the heterogeneity (in terms of the number of missing entries of clinical information) affects the performance of each aggregation method. To do so, we use a fixed batch size of 50 and 100 epochs to train each FCAE with corruption rates of 10%, 30% and 60% for 300 iterations. Figure 1 shows the performance evaluations between FedAvg and PW and the reconstruction estimations for non-missing scores (A1). We observe that FedAvg achieves up to 2.7% lower reconstruction error than PW with a 10% corruption ratio. However, PW reaches 16.4% and 5% lower errors than FedAvg with a 30% and 60% corruption ratio, respectively. This experiment suggests that a weighted average can be effective with subtle variations in training data across centers, but when variability in the input data is considered into the aggregation, more accurate imputations and reconstructions may be obtained with highly heterogeneous inputs.

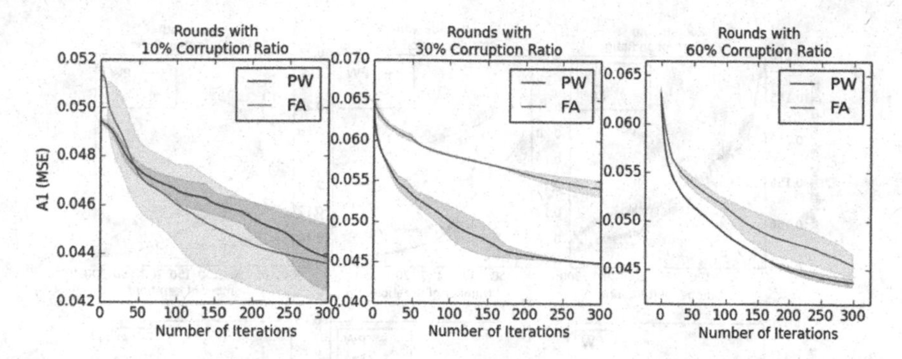

Fig. 1. Performance of FedAvg and PW aggregation algorithms based on the reconstruction error of known values (A1) with an increasing number of missing data in the training dataset.

Table 2. Summary of the performance of the central model and two aggregation algorithms based on known values (A1) on the test set

Corruption Ratio	FA	PW	Central
10%	0.044 ± 0.002	0.044 ± 0.001	0.192 ± 0.000
30%	0.054 ± 0.002	0.045 ± 0.004	0.183 ± 0.000
60%	0.045 ± 0.004	0.044 ± 0.004	0.166 ± 0.000

As a point of comparison for model generalization with distributed models, we trained a central model with the pooled dataset (test set excluded). Table 2 summarizes the A1 (MSE) scores obtained on the hold-out test set. Interestingly, these results begin to demonstrate an advantage in using distributed algorithms as we observe lower A1 scores on the reconstruction of non-missing clinical scores compared to those values obtained with a centralized learning setting.

4.2 Effect of Number of Missing Values During Evaluation

We evaluate the reconstruction performance of each aggregation method with additional missing scores. We set the experiment with the same training hyper-parameters as above. This time, we introduce additional missing scores into the test set with a 10% and 20% chances for a given score to be missing and compute its A2 error between the estimation and ground truth. Given this setup, it will allow us to explore the scenario when the imputation is implemented in new sites with different levels of missing patient records. Figure 2 shows the performance evaluations based on the A2 (MSE). As expected, the accuracy of the reconstruction A2 is affected by the number of missing scores injected. More specifically, FedAvg shows 1% and 4% lower error than PW with 10% and 20% missing ratio in test, respectively, and 10% corruption ratio. This suggests that FedAvg is a more robust method for the aggregation of homogeneous data. On

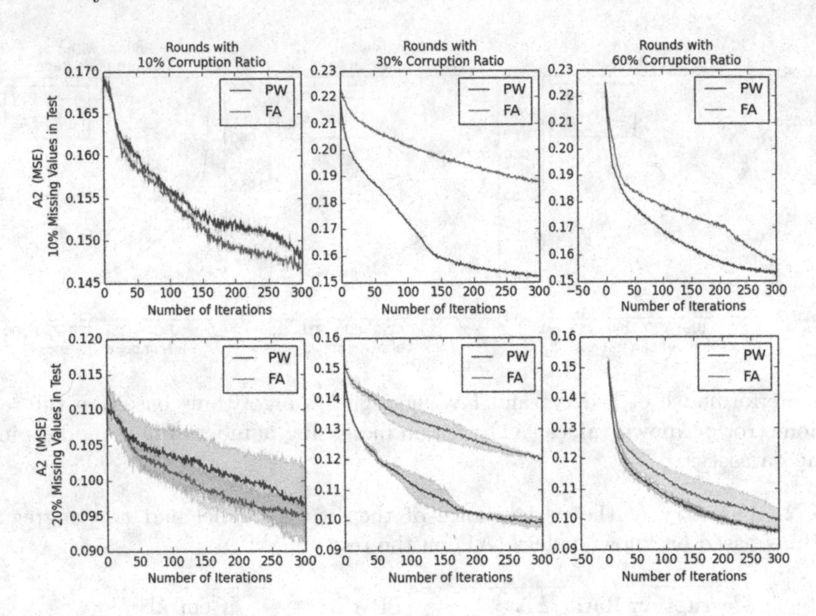

Fig. 2. Performance of FedAvg and PW aggregation algorithms based on the reconstruction error of missing values (A2) with an increasing number of missing values in the testing dataset.

the contrary, PW reached up to 19% lower reconstruction error than FedAvg with additional heterogeneity in either the training or test sets.

Further evaluations with a central model trained with the pooled dataset were conducted to measure the model generalization based on the reconstruction estimations for missing values. Table 3 summarizes the A2 (MSE) scores of the central model and each aggregation algorithm with 10% and 20% missing values in the test set. We obtained lower A2 scores on central models as more data is available during training for the estimation of missing scores. However, this effect was pronounced when we introduced 10% missing values into the test set. Alternatively, distributed learning improves performance over sites with higher

Table 3. Summary of the performance of the central model and two aggregation algorithms based on the missing values (A2) on the test set

Miss Ratio	Corruption Ratio	FA	PW	Central
10%	10%	0.146 ± 0.006	0.148 ± 0.005	0.117 ± 0.000
	30%	0.187 ± 0.008	0.152 ± 0.017	0.117 ± 0.001
	60%	0.156 ± 0.012	0.152 ± 0.013	0.118 ± 0.002
20%	10%	0.095 ± 0.004	0.096 ± 0.003	0.164 ± 0.000
	30%	0.120 ± 0.008	0.098 ± 0.011	0.167 ± 0.000
	60%	0.099 ± 0.010	0.095 ± 0.010	0.165 ± 0.001

levels of missing patient scores (20% random missing values). These results are consistent with the findings of Tuladhar *et al.* in [20] and suggest that better generalization can be obtained with distributed models.

5 Discussion and Conclusion

We compared two aggregation algorithms for distributed learning environments and demonstrated that medical and research centers can embrace collaborate learning to enrich estimations of statistical analysis for the severity and progression of Parkinson's disease. To the best of our knowledge, this is the first work that provides a comparative evaluation of the Federated Averaging and the Precision-weighted Federated Learning aggregation algorithms in the demonstrated domain.

The first point of discussion is that regardless of the aggregation algorithm used, we observe a significant benefit by sharing multi-center clinical data for collaborative model training. We showed that training independent distributed models with information from PD patients can increase the model's generalizability on a hold-out test set without transferring patient's records to a central data store. Notwithstanding, both aggregation algorithms remain vulnerable to inference attacks, therefore, stronger privacy guarantees are needed to protect the information transferred across sites. One solution is using secure protocols [3] or differential-privacy guarantees [6,14] to ensure that data is transferred between clients and servers safely. Alternatively, we demonstrated that distributed data not only augments the size and variety of the global training set, but also it increases clinical utility. For example, the prediction of the progression and trajectory of the disease can be achieve with less biased decisions than models trained with data in single institutes, leading to more effective treatments or preventive strategies.

An important outcome from the present study is the evaluation of two distributed aggregation algorithms for the imputation and reconstruction of missing scores in PD clinical assessments. Our study indicated that with the lowest level of corruption introduced into the training data, FedAvg can achieve better generalization on a hold-out test set (based on A1 and A2 MSE) with a weighted average, despite the subtle variations in the training data. These results can be explained by the fact that PD patients exhibit high variations in disease patterns and when the number of incomplete responses rise the heterogeneity of the data increases artificially. However, this effect is not pronounced as the corruption rate increases. With less information available to the model to perform the reconstruction, PW seems to be better suited for the reconstruction task.

In addition, it is important to highlight that a randomly initialization for FCAEs was employed in our experiments. Perhaps with a better initialization strategy we could obtain better estimations, especially in the reconstruction errors for (A2). Despite this, we observed that machine/deep learning models were able to leverage the condense information in the clinical scores and factors to provide meaningful and accurate estimations that can be used to perform imputation and reconstruction tasks with acceptable clinical outcomes. Future work may investigate methods for combining information from local datasets, such as cyclical weight transfer [4], split learning [21], or transfer learning [17] as these offer performance improvements with small local training sets.

In conclusion, we present a comparative analysis of the performance of two aggregation algorithms for distributed learning: Federated Averaging (FedAvg) and PW Federated Learning. The task explored here is data imputation and reconstruction of Parkinson's disease clinical questionnaires. We built upon the work of Peralta *et al.* and evaluate the reconstruction performance with Fully-Connected Autoencoders operating in a distributed environment. The results in this study demonstrated that FedAvg is effective in estimating the reconstruction of data with subtle differences across centers, but PW poses a better choice when data is highly heterogeneous.

Acknowledgement. The data that support the findings of the study are publicly available at www.ppmi-info.org. PPMI—a public-private partnership—is funded by the Michael J. Fox Foundation for Parkinson's Research and funding partners, including AbbVie, Avid, Biogen, Bristol-Myers Squibb, Covance, GE Healthcare, Genentech, GlaxoSmithKline, Lilly, Lundbeck, Merck, Meso Scale Discovery, Pfizer, Piramal, Roche, Sanofi Genzyme, Servier, Teva, and UCB.

References

1. Allison, P.D.: Missing Data. Sage publications, Thousand Oaks (2001)
2. Azur, M.J., Stuart, E.A., Frangakis, C., Leaf, P.J.: Multiple imputation by chained equations: what is it and how does it work? Int. J. Methods Psychiatr. Res. **20**(1), 40–49 (2011)
3. Bonawitz, K., et al.: Practical secure aggregation for federated learning on user-held data. arXiv preprint arXiv:1611.04482 (2016)
4. Chang, K., et al.: Distributed deep learning networks among institutions for medical imaging. J. Am. Med. Inform. Assoc. **25**(8), 945–954 (2018)
5. Denis, A., Mergaert, L., Fostier, C., Cleemput, I., Simoens, S.: A comparative study of European rare disease and orphan drug markets. Health Policy **97**(2–3), 173–179 (2010)
6. Dwork, C., Roth, A., et al.: The algorithmic foundations of differential privacy. Found. Trends Theor. Comput. Sci. **9**(3–4), 211–407 (2014)
7. Efron, B.: Missing data, imputation, and the bootstrap. J. Am. Stat. Assoc. **89**(426), 463–475 (1994)
8. Ioannidis, J.P., Patsopoulos, N.A., Evangelou, E.: Heterogeneity in meta-analyses of genome-wide association investigations. PLoS ONE **2**(9), e841 (2007)
9. Jankovic, J.: Parkinson's disease: clinical features and diagnosis. J. Neurol. Neurosurg. Psychiatry **79**(4), 368–376 (2008)
10. Lachin, J.M.: Fallacies of last observation carried forward analyses. Clin. Trials **13**(2), 161–168 (2016)
11. Marek, K., et al.: The parkinson progression marker initiative (PPMI). Prog. Neurobiol. **95**(4), 629–635 (2011)
12. McGhee, D.J., Royle, P.L., Thompson, P.A., Wright, D.E., Zajicek, J.P., Counsell, C.E.: A systematic review of biomarkers for disease progression in Parkinson's disease. BMC Neurol. **13**(1), 1–13 (2013)

13. McMahan, B., Moore, E., Ramage, D., Hampson, S., y Arcas, B.A.: Communication-efficient learning of deep networks from decentralized data. In: Artificial Intelligence and Statistics, pp. 1273–1282. PMLR (2017)
14. Melis, L., Song, C., De Cristofaro, E., Shmatikov, V.: Exploiting unintended feature leakage in collaborative learning. In: 2019 IEEE Symposium on Security and Privacy (SP), pp. 691–706. IEEE (2019)
15. Peralta, M., Jannin, P., Haegelen, C., Baxter, J.S.: Data imputation and compression for Parkinson's disease clinical questionnaires. Artif. Intell. Med. **114**, 102051 (2021)
16. Price, W.N., Cohen, I.G.: Privacy in the age of medical big data. Nat. Med. **25**(1), 37–43 (2019)
17. Raghu, M., Zhang, C., Kleinberg, J., Bengio, S.: Transfusion: understanding transfer learning for medical imaging. arXiv preprint arXiv:1902.07208 (2019)
18. Reyes, J., Di Jorio, L., Low-Kam, C., Kersten-Oertel, M.: Precision-weighted federated learning. arXiv preprint arXiv:2107.09627 (2021)
19. Sterne, J.A., et al.: Multiple imputation for missing data in epidemiological and clinical research: potential and pitfalls. Bmj 338 (2009)
20. Tuladhar, A., Gill, S., Ismail, Z., Forkert, N.D., Initiative, A.D.N., et al.: Building machine learning models without sharing patient data: a simulation-based analysis of distributed learning by ensembling. J. Biomed. Inform. **106**, 103424 (2020)
21. Vepakomma, P., Gupta, O., Swedish, T., Raskar, R.: Split learning for health: Distributed deep learning without sharing raw patient data. arXiv preprint arXiv:1812.00564 (2018)
22. Voon, V., Fox, S.H.: Medication-related impulse control and repetitive behaviors in parkinson disease. Arch. Neurol. **64**(8), 1089–1096 (2007)
23. Zhang, X., et al.: Data-driven subtyping of Parkinson's disease using longitudinal clinical records: a cohort study. Sci. Rep. **9**(1), 1–12 (2019)

Defending Medical Image Diagnostics Against Privacy Attacks Using Generative Methods: Application to Retinal Diagnostics

William Paul[1(✉)], Yinzhi Cao[2], Miaomiao Zhang[3], and Phil Burlina[1,2,4]

[1] Applied Physics Laboratory, Johns Hopkins University, Laurel, MD, USA
william.paul@jhuapl.edu

[2] Department of Computer Science, Johns Hopkins University, Laurel, USA

[3] Department of Electrical Engineering, University of Virginia, Charlottesville, USA

[4] Malone Center for Engineering in Healthcare, Johns Hopkins University, Laurel, USA

Abstract. Machine learning (ML) models used in medical imaging diagnostics can be vulnerable to a variety of privacy attacks, including *membership inference attacks*, that lead to violations of regulations governing the use of medical data and threaten to compromise their effective deployment in the clinic. In contrast to most recent work in privacy-aware ML that has been focused on model alteration and post-processing steps, we propose here a novel and complementary scheme that enhances the security of medical data by controlling the data sharing process. We develop and evaluate a privacy defense protocol based on using a generative adversarial network (GAN) that allows a *medical data sourcer* (e.g. a hospital) to provide an external agent (*a modeler*) a proxy dataset synthesized from the original images, so that the resulting diagnostic systems made available to *model consumers* is rendered resilient to privacy *attackers*. We validate the proposed method on retinal diagnostics AI used for diabetic retinopathy that bears the risk of possibly leaking private information. To incorporate concerns of both privacy advocates and modelers, we introduce a metric to evaluate privacy and utility performance in combination, and demonstrate, using these novel and classical metrics, that our approach, by itself or in conjunction with other defenses, provides state of the art (SOTA) performance for defending against privacy attacks.

Keywords: Medical data privacy · Generative models · Retinal diagnostics

1 Introduction

There has been a recent proliferation of artificial intelligence (AI) and machine learning (ML) applications being developed and proposed for deployment in

C. Oyarzun Laura et al. (Eds.): CLIP/DCL/LL-COVID/PPML 2021, LNCS 12969, pp. 174–187, 2021.
https://doi.org/10.1007/978-3-030-90874-4_17

various tasks ranging from vision [13,15,36] to natural language processing and speech [11,25,26,32]. However, ensuring guarantees of privacy for the data used for training those applications and for medical and retinal AI diagnostics [3,4, 7,12,24,29–31] is shaping up as an open impediment to deployment. For the purposes of this work, we focus on ensuring privacy when a trained classification model (classifier) created by a *modeler* is accessible to an attacker, allowing them to acquire information about individuals whose data was used in the training process. The attacker may use the model to infer private attributes (e.g. age or co-morbidities) about a specific person, in what is called an *attribute inference attack* [9], or to determine if an individual's data was used for training, in what is termed a *membership inference attack* [27,28,34], which is the focus of this study. Notably, we focus in this work on the case of the data being stored in a central location under the provenance of a trusted agent called a *medical data sourcer*, acting under an institutional review board that provides for an individual's privacy. Although we focus on having a single data sourcer for this work, we believe this work can be extended to the federated scenario [18,33] where there are multiple sources of data.

This work focuses on how access to the data can be controlled by the medical data sourcer, namely creating new data points that should not contain the true identities of the original individuals that can safely be passed to modelers. This work evaluates the effectiveness of synthetic data for privacy, on retinal imagery collected for the task of diabetic retinopathy, alongside other defenses that affect the classifier directly. Due to the degree of data access being controlled, in terms of how many private data points are shared, there is a question of how to capture the typical trade-off between the privacy conferred by the model as well as the performance. Consequently this work introduces an additional metric that attempts to capture the trade off in a single measure, allowing for negotiation about the level of access between the data sourcer and modeler.

The unique and salient contributions of this work include:

1. We develop a novel strategy for privacy defense based on rejecting potentially vulnerable samples from generative models, which only depends on the source of the data, and requires no change in procedure in training. We believe that this approach can be used more broadly for other image-based medical diagnostics and other image classification tasks beyond healthcare.
2. We propose a novel metric, called P1-score, to measure the trade-off between utility and privacy, so that privacy advocates and modelers can view both concerns together in contrast to most methods which only look at each individually.

2 Background

Being able to infer if an individual's data was used for training has a variety of possibly severe implications with regard to privacy violations, generically leading to the discovery – via conflation with other public information – of private information on the individual, or information on the medical data sourcer, or both.

For individuals, leaking knowledge about membership may cause an attacker to realize that a relationship between the individual and the healthcare entity exists, which is a problem as the attacker may have collected additional metadata about the individual that could be a focus for further attacks. Membership information could also be part of a linkage attack, as both the image and whether it was used for training may imply additional private information about the individual. Additionally, the individual may not want the relationship itself to be known, and violating that desire may erode the trust placed in the healthcare entity. Finally, membership inference attacks can be considered gateways to even more critical attacks termed *training data reconstruction attacks* which allow for recovery of training data from trained models [6].

As the classifier is the main mode of leakage, most defenses [1,16,20,27,34] focus on augmenting the model either during training or inference. These schemes all rely on the modeler to actually implement these measures on the given data, and the medical data sourcer trusting the modeler or attempting to audit the classifier directly. However, the incentives for the modeler typically favor pure performance over other concerns, potentially causing privacy to be an afterthought. Most ways to influence these incentives are to enforce compliance through regulation such as Health Insurance Portability and Accountability Act (HIPAA), rather than explicitly rewarding privacy aware models. Moreover, unlike more traditional settings in security, typically the only means for the medical data sourcer to audit the trained model is by the same attacks an attacker would use, either emboldening modelers who maliciously ignore regulation to improve performance, as they could pass a different model from what is actually used to auditors, or not catching ignorant modelers who use bad practices in training.

3 Prior Work

For defending implementations of classifiers against membership inference attacks, there exists methods with both empirical and theoretical successes. For empirical methods, most existing work has focused on regularization of the model during training or inference time. One approach is to reduce overfitting on the training set, as the disparity between training and testing data could arguably be the primary source for identifying training data points. Techniques such as dropout [34], which randomly drops part of an activation layer during training, L2 Regularization [27], which encourages the smaller weights to help prevent overfitting, and MMD+Mixup [20]. MMD+Mixup combines Mixup, a data augmentation technique that samples random linear interpolations between two data points for both the image and the label, and an MMD regularization term that tries to match the average probability vector between training and validation data.

There are also methods that use an surrogate adversary to defend against membership inference attacks. For example, Nasr et al. [23] turns training into a minimax procedure, where the classifier is trying to both classify the training

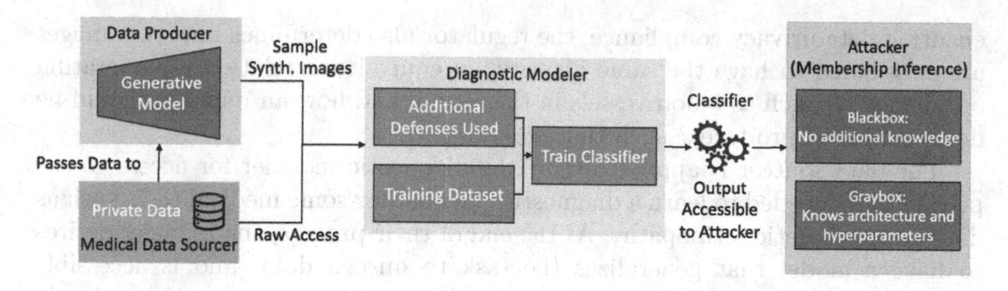

Fig. 1. Overview of our proposed approach and division of duties among agents.

data correctly while fooling an adversary, which in turn is trained to distinguish between training and reference data. Another adversarial method called MemGuard [16] is notable, as it only affects inference of the trained model and is utility preserving. It trains a surrogate to determine membership using the trained model's logits as input like [23], and minimally perturbs the trained model's logits to fool the surrogate while keeping the same argmax or predicted label. Finally, there are methods that use Generative Adversarial Networks (GANs) to create synthetic data that does not contain identifying information [17], where we focus on curating the generated imagery directly. However, trusting external collaborators with training these models without evaluating the samples can be perilous, as these collaborators could exfiltrate the private data out using the GAN as [22] shows. For theoretical methods, the most prominent example is differential privacy (DP), notably DP-Adam [1], which primarily adds noise to gradients to eliminate that minor perturbations in the gradient that could leak the identity of the data point.

Although most methods that are used to address privacy demonstrate empirical successes, only a subset of these methods offer theoretical guarantees for their performance. Most methods with no guarantees, notably dropout, L2 regularization, and Mixup, were lifted from work studying generalization. Adversarial methods in the former category typically use a specific adversary to defend against attacks, and a stronger adversary may nullify such defenses. Turning training into an adversarial game as with [23] also increases the complexity of training classifiers. For methods with theoretical backing, using differential privacy is, in general beyond a notion of a privacy budget, opaque to the modeler, causing issues such as inducing disparity with respect to subpopulations [2].

4 Methodology

First, we describe the roles the medical data sourcer, the data producer, the diagnostic modeler, and the attacker play in our methodology, with an overview given in Fig. 1.

The medical data sourcer is a healthcare entity, such as a hospital, that both has amassed a collection of valuable and private data from individuals and is bound by an institutional review board or other privacy regulator. Alongside ensuring data privacy compliance, the regulator also determines how two images are considered to have the same identity, an equivalence relation $\sim_I$ operating on biometrics such as blood vessels in the retina, and how an image x should be de-identified to produce $\hat{x}$ such that $x \not\sim_I \hat{x}$.

The data sourcer is approached by the diagnostic modeler for access to the private data, needed to learn a diagnostic classifier for some medical task, such as diagnosing diabetic retinopathy. At the end of their process, the modeler desires to have a model that generalizes the task to unseen data, and is accessible in some way by model consumers and attackers. Although the modeler simply needs the private data to create a classifier that works on novel data, common schemes to de-identify typically harm generalization without consideration by the modeler, as novel images are unlikely to have de-identifying alterations like masks. Consequently, most modelers are instead authorized to have raw access to the private data, complicating the liability of the medical data sourcer.

In the case where the data sourcer wants to mitigate privacy leaks from sharing too much data, they turn to a data producer to de-identify images, as they might not have the technical expertise themselves. More traditional methods of doing this primarily involve censoring areas of the image to hide identifiable information as well as hiding metadata. However, this does not preserve the realism of the images, harming performance of the resulting model. Consequently, the data producer can instead use other methods, such as generative adversarial networks, to defend against attackers of the final model attempting to determine the membership of data points, both detailed next.

4.1 Threat Model

We define our threat model for classifiers using similar notation as in [5]:

Definition 1 (Membership Inference Attack). $D = \{(x_1, y_1), \ldots, (x_n, y_n)\}$ *denotes a dataset, where* x_i *being images and* y_i *the labels, sampled from* $p(x, y)$*, acquired by the modeler from either the data producer or the data sourcer. This dataset, not known by the attacker, is used to train a classifier* F *using training settings* S*, i.e. hyperparameters, optimizers, and architecture used but not any defenses used. Given access to* $p(x, y)$*, the goal of the attacker is to choose* (x^*, y^*) *in the support of* $p(x, y)$ *such that* $x^* \sim_I x_i$*, for an equivalence relation* $\sim_I$ *and any* $i \in [n]$*, under one of two settings:*

- **Blackbox setting:** *where the attacker has access to the model as an oracle* $G(x)$*, i.e. whereby the attacker is able to take an input datum and get the probabilities over* y *produced by* $F(x)$*.*
- **Graybox setting:** *which allows the adversary access to both* $G(x)$ *and* S*.*

and the following assumption:

Assumption 1 (Locality of Inference). *If a data point* x *is used to train a classifier* F*, then for any data point* $\tilde{x} \neq x$*,* $x \not\sim_I \tilde{x}$ *implies that it is not possible for an attacker to find* x^* *using any level of access to* F *such that* $\tilde{x} \sim_I x^*$*. In* *other words, a data point used for training F does not leak data points that have different identities.*

As we are explicitly not focused on reconstruction attacks where the attacker could learn $p(x, y)$ directly, the attacker is assumed to have ground truth knowledge of it. $\sim_I$ can be interpreted as comparing the identity of the two images, and an equivalence means that they have the same identity.

4.2 Approach for Data Producer to Defend Privacy

In order to produce data points that are usable by the modeler, there are three desirable properties for the generated data: (a.) to preserve the original task, i.e. for classification this means being aligned with a certain class, (b.) to be realistic so that the classifier trained on this dataset can generalize, and (c.) to ensure that the generated data is not equivalent to the original data in the sense of $\sim_I$, which by Assumption 1 is sufficient to preserve privacy. (a.) and (b.) can be effectively resolved by resampling from $p(x, y)$, the true data distribution. However, the original way to sample from this distribution is to acquire data from individuals, where privacy concerns arise. Consequently, the data producer wants to construct a surrogate distribution $\tilde{p}(x, y)$ from the private data that should mimic the true distribution, but without further interaction with any individuals.

For this work, we focus on the data producer using Generative Adversarial Networks (GANs) [10], namely StyleGAN2-ADA [19], to generate this synthetic data. The most desirable properties of GANs are their ability to create realistic data that should conform to the true data, which we leverage here. [19] also includes provisions for training on smaller datasets, enhancing its usefulness on medical imagery. To model $p(x, y)$, the generator and discriminator are made conditional on the label y. The data producer can then fix the desired label and sample from the generator to create the data to pass to the modeler. See Figs. 2 and 3 for real and synthetic data examples.

To fully satisfy (c.), influencing the sampling to move away from the original points is done by rejecting samples that are equivalent with respect to $\sim_I$. However, it is difficult to fully specify $\sim_I$ in a mathematical form, beyond simple, incomplete measures such as those based on a threshold on the L^2 distance in the raw pixel space, which we use in this work. In real world settings, $\sim_I$ will need to be carefully defined using known indicators of identity, i.e., blood vasculature, in order to ensure patient privacy. Consequently, samples are ensured not to immediately return members of the training dataset, and we argue there exist realistic samples that are not identifiable as our generator is continuous. Namely, even if most of the probability support is on the original dataset, there exist interpolations between these points that are sampled and made realistic by GAN training.

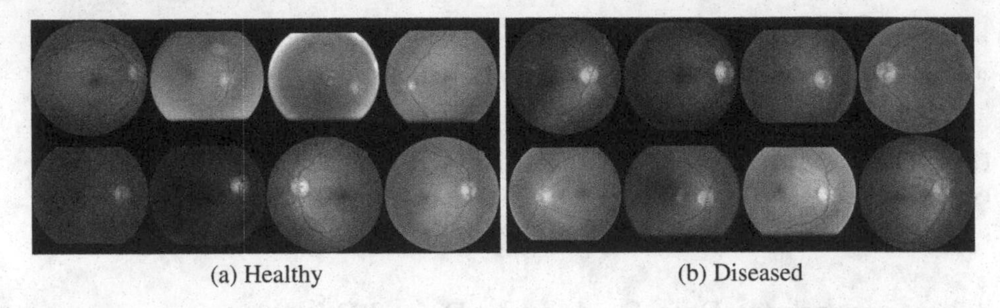

Fig. 2. Example real retinal images. Labels denote the severity of the diabetic retinopathy from 0 to 4, where 0,1 are taken to be healthy and 2,3,4 to be diseased.

4.3 Novel Metric Balancing Utility and Privacy

As the medical data sourcer determines what degree of access the modeler has to the original data, metrics that combine the utility and the privacy leakage of the final model are needed for determining the level of access. Indeed, we posit that the field of security and privacy would benefit from the design of more effective metrics that capture possible tradeoffs between utility and privacy: encouraging higher accuracy for accomplishing the task, i.e. utility, and attenuating accuracy of the attackers in breaching privacy of the diagnostics model, i.e. privacy. We propose here to use a novel metric modeled after the popular F1 score, and which would measure the harmonic mean of the classifier's accuracy and the attack's error rate:

$$P1(D)_{Attack} = 2 * \frac{(Acc_{Task,D}) * (1 - Acc_{Attack,D})}{(Acc_{Task,D}) + (1 - Acc_{Attack,D})} \tag{1}$$

where $Acc_{Task,D}$ denotes the accuracy of the defended model, and $Acc_{Attack,D}$ denotes the accuracy of the attacker on the defended model.

5 Experiments

As this work is focused on the perspective of the medical data sourcer worried about privacy violations, we focus on the worst case scenario of overfitting the trained classifiers as much as possible [20]. To mimic the varying levels of data access the data sourcer can provide to control privacy leaks, we combine the synthetic data and real data for the modeler's training set at different proportions while keeping the overall size constant.

We use ResNet50, initialized to pretrained ImageNet weights, as our diagnostic classifier, and fix the number of epochs trained to 15 for all experiments, well in excess of what achieves the best accuracy and overfitting. The Adam optimizer is used with a batch size of 64, a learning rate of $5 * 10^{-5}$, with $\beta_1 = 0.9$ and $\beta_2 = 0.999$, and no validation set is used for the baseline model. The training set for the classifier is used to train the GAN, with default training settings

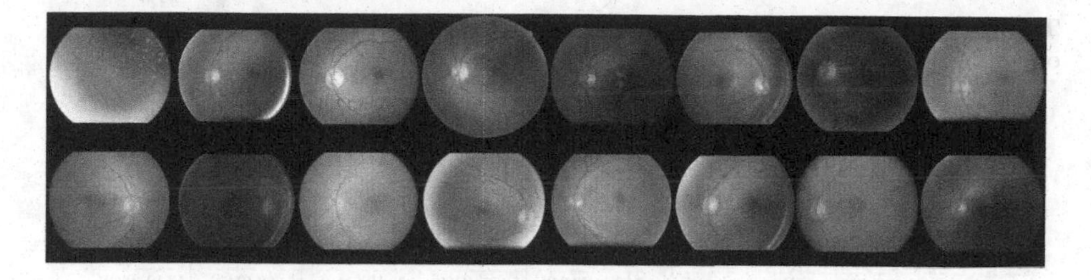

Fig. 3. Example synthetic retinal images from data producer.

besides a batch size of 32, learning rate of $2.5 * 10^{-3}$, and R1 penalty of 1. For each mixture of synthetic data and real data, we evaluate both training with no defenses and training with MMD+Mixup. We partition the training dataset into 80% training and 20% validation. The interpolation weight for Mixup is sampled from a $beta(\lambda, \lambda)$ where λ is between $\{0, 10\}$. $\lambda = 5$ was found to perform the best with regards to our combined metric, and is shown in both tables. On real data only, we also evaluate on MemGuard, with the same settings from the original work. The task and attack accuracy are evaluated with respect to the modeler's datasets.

For the adversary attempting to determine the membership of training images, they use two attacks from [34], loss-threshold (Loss-Thre) attacks or label-only attacks. Loss threshold attacks involve the attacker training their own model, called a shadow model, on data they themselves have collected. This shadow model should match the original classifier as much as possible, and this model in the Graybox setting is trained using the same architecture (pretrained ResNet50) and hyperparameters as for our diagnostic classifier in the previous paragraph. For the Blackbox setting, as the attacker does not know the true training settings, we use a pretrained VGG16 network instead and change the number of epochs trained to 9. The remaining configuration is the same as for the graybox and diagnostic classifier.

Once this shadow model is trained, the attacker can use it to determine statistics about the training and test set used to train the shadow model, and assume these statistics also hold for the classifier being attacked. Namely, the attacker determines the average loss (cross-entropy for these experiments) of the shadow model's training set, so then any image having a cross entropy with respect to the classifier smaller than this average is a member of the classifier's training dataset. The label-only attack is much simpler, and does not require a shadow model be constructed and thus is independent of the attach setting. This attack simply predicts membership in the classifier's training dataset for a target image if the classifier produces the correct answer for this image, i.e., training set members are more likely to be correct compared to data that was not trained on.

Table 1. Task and attack accuracies (%) of various defenses with different percentages of synthetic data. 95% CIs are in parenthesises, and numbers in bold are the best.

	Raw data access %	Defense	$Acc_{Task,D}$	$Acc_{Attack,D}$			
				Blackbox		Graybox	
				Loss-Thre	Label-Only	Loss-Thre	Label-Only
Synthetic data only	0%	No defense	68.77 (0.91)	49.84 (0.69)	49.90 (0.69)	**49.85 (0.69)**	49.95 (0.69)
		MMD+Mixup	73.30 (0.87)	**49.67 (0.69)**	**49.79 (0.69)**	49.99 (0.69)	**49.79 (0.69)**
Synthetic/real mixture	25%	No defense	73.54 (0.86)	53.35 (0.69)	53.10 (0.69)	54.98 (0.69)	53.10 (0.69)
		MMD+Mixup	74.10 (0.86)	52.65 (0.69)	52.17 (0.69)	50.05 (0.69)	52.17 (0.69)
	50%	No defense	72.95 (0.87)	57.49 (0.69)	56.70 (0.69)	60.31 (0.68)	56.70 (0.69)
		MMD+Mixup	73.80 (0.86)	51.91 (0.69)	52.86 (0.69)	50.08 (0.69)	52.86 (0.69)
	75%	No defense	75.14 (0.85)	60.43 (0.68)	59.21 (0.68)	66.03 (0.66)	59.21 (0.68)
		MMD+Mixup	74.75 (0.85)	54.52 (0.69)	55.41 (0.69)	50.01 (0.69)	55.41 (0.69)
Real data only	100%	No defense (baseline)	73.24 (0.87)	64.25 (0.66)	62.70 (0.67)	66.64 (0.65)	62.70 (0.67)
		MMD+Mixup	**75.52 (0.84)**	61.80 (0.67)	59.23 (0.68)	50.08 (0.69)	59.23 (0.68)
		Memguard	73.24 (0.87)	63.87 (0.67)	62.70 (0.67)	63.87 (0.67)	62.70 (0.67)

5.1 Dataset

For our experiments, we use the EyePACs dataset from Kaggle [8], originally used for a Diabetic Retinopathy Detection challenge. The dataset includes 88,703 high-resolution retina images taken under a variety of imaging conditions and each image has a label ranging from 0 to 4, representing the presence and severity of diabetic retinopathy.

We select 10,000 random images each for modeler's training and testing set and the attacker's training and testing set, all disjoint. The images are cropped to the boundary of the fundus, and resized to 256 by 256 pixels. The GAN is trained on this data, and the classifier has additional processing for contrast normalization. The GAN used is Stylegan2-ADA [19] using Adaptive Data Augmentation with a minibatch of 32, minibatch standard deviation layer with 8 samples, a feature map multiplier of 1, learning rate of 0.0025, gamma of 1, and 8 layers in the mapping network. The rest of the configuration remains the same from the original work.

To reject samples that are identifiable, the threshold we use is the minimum L^2 distance in pixel space between images in the training dataset, as we know that images in the training dataset are distinct. Consequently, a synthetic image that is a distance smaller than this threshold away from any training image is taken to have the same identity and rejected. The actual threshold computed was 2328.96 for the dataset we use, where the minimum pixel value is 0 and the maximum 255 for computing the threshold. Consequently, none of the synthetic images we generated are within this threshold to a point, and, as the threshold is greater than zero, not equivalent to any training image.

When evaluating the quality of the generated images, we use the Frechet Inception Distance (FID) [14], a standard metric in GAN literature that computes the distance between real image representations and synthetic image representations taken from a layer of the Inception network. More specifically, each set of representations is assumed to be Gaussian, so the Frechet distance between

Table 2. $P1(D)_{Attack} \in [0, 100]$ where larger is better for various settings and defenses.

	Raw data access %	Defense	$P1(D)_{Attack}$			
			Blackbox		Graybox	
			Loss-Thre	Label-Only	Loss-Thre	Label-Only
Synthetic data only	0%	No defense	58.01 (0.63)	57.97 (0.63)	58.00 (0.63)	57.94 (0.63)
		MMD+Mixup	**59.68 (0.62)**	**59.60 (0.62)**	59.46 (0.62)	**59.60 (0.62)**
Synthetic/real mixture	25%	No defense	57.09 (0.61)	57.27 (0.61)	55.85 (0.61)	57.27 (0.61)
		MMD+Mixup	57.78 (0.61)	58.14 (0.61)	59.67 (0.61)	58.14 (0.61)
	50%	No defense	53.72 (0.61)	54.34 (0.61)	51.41 (0.60)	54.34 (0.61)
		MMD+Mixup	58.23 (0.61)	57.53 (0.61)	59.56 (0.61)	57.53 (0.61)
	75%	No defense	51.84 (0.60)	52.88 (0.60)	46.79 (0.58)	52.88 (0.60)
		MMD+Mixup	56.55 (0.61)	55.86 (0.61)	59.91 (0.61)	55.86 (0.61)
Real data only	100%	No defense (baseline)	48.05 (0.59)	49.43 (0.60)	45.84 (0.59)	49.43 (0.60)
		MMD+Mixup	50.74 (0.59)	52.95 (0.60)	**60.11 (0.61)**	52.95 (0.60)
		Memguard	48.39 (0.59)	49.43 (0.60)	48.39 (0.59)	49.43 (0.60)

the two distributions can be computed. The FID of the generator we used for this work is 4.12. In addition, we asked three highly trained retinal specialists to critically assess images in Fig. 3 for vessels' realism, with two questions: how many look real? (answers: 16, 12, and 7) and how many have potential vasculature issues? (answers: 1, 4, 9). While empirical results show the potential of our method for classification, these clinical assessments suggest future work incorporating a loss term to regularize the structure of the resulting blood vessel branching and to increase their realism which would make the generated images useful for example as private datasets that are also adequate for resident training. From our qualitative evaluation, we believe that our generator produces images of an acceptable quality for training, and assign these images the labels that were used to generate them (as our generator is conditional on the class label).

5.2 Results

In Table 1, we see that using only synthetic data for training completely defeats the attacker (attack probability is close to chance), at the cost of decreased accuracy. Using MMD+Mixup on synthetic data only improves the task accuracy to be on par with the original classifier, and close to MMD+Mixup with only real data. Due to its utility preserving nature, MemGuard does not outperform MMD+Mixup for attack accuracy nor task accuracy. For most attacks, the attack accuracy with no defenses increases roughly linearly with the amount of real data in the training dataset. With the MMD+Mixup defense, the relationship is less clear, possibly due to the dataset partitioning. For the combined metric $P1(D)_{Attack}$ in Table 2, we see that synthetic data only with MMD+Mixup is the best of breed against all attacks except the Graybox loss-threshold attack, for which the best performance is obtained by the MMD+Mixup on real data only due to its higher task accuracy, but the difference between that and the second best method which uses 100% synthetic data with MMD+Mixup is barely significant.

6 Discussion and Limitations

Though we have demonstrated that our synthetic data can be used to recover similar accuracies as training with real data in this case, in general GANs have issues with being able to fully represent all modalities of the training dataset. Improvements in generative modeling are still being made that should address this deficiency, but in the meantime, we note that the data sourcer can give raw access to data necessary for the classifier to perform well. We see from Table 1 that the attack accuracy for classifiers with no defenses increases linearly with the amount of real data included, suggesting that each individual data point trained on only violates the privacy of itself and Assumption 1 holds. Consequently, the sourcer can still protect most of the population while giving the modeler access to data that can be instrumental to model performance.

Our introduced $P1(D)_{Attack}$ metric uses the overall accuracy as the primary measure of utility, but other measures such as F1 score, precision, or area under the receiver operating characteristic curve (AUC) could be used instead. The accuracy in the formulation of the metric can be substituted by these measures directly, but one concern that becomes more apparent is how to normalize changes in each individual metric such that the actual numerical value matches user preferences. For example, the user may prefer a 3% increase in accuracy or a 0.05 increase in the F1 score over a 7% decrease in the attack accuracy, but the current $P1$ formulation uses a fixed, equal weighting between utility and privacy. A simple way to address this is to allow the user to set this weighting, but more investigation will be needed for how well this can match preferences.

Although the equivalence relation $\sim_I$ we use for this work was effective in defeating adversaries, more sophisticated measures can help situations where an image is far away in pixel space but still retains the same identity, such as when artifacts are present or the overall brightness is different. For retinal imagery, these can be based on biometrics such as the retinal vasculature, which can be extracted via vessel segmentation techniques [21] and compared between images. For general medical imagery, more robust measures of perceptual similarity can be used, such as structural similarity index measure (SSIM), or metrics such as LPIPS [35] that use the intermediate layers of deep networks.

Finally, a limitation of our work is not comparing directly to differential privacy (DP), which should be done in future work. Additionally, Assumption 1 is reliant on the definition of $\sim_I$ to assume privacy, and, as discussed in the previous paragraph, there could be uncertainty about whether $\sim_I$ captures the true identity. One interesting direction to extend this work is to treat sampling and rejecting synthetic data as a random mechanism in the same way DP treats computing the gradient as a random mechanism, and use DP's notion of a privacy budget in place of the assumption.

7 Conclusion

We propose a novel approach using generative methods to defend membership inference attacks of retinal diagnostics. Our evaluation shows that, used alone or in combination of SOTA defenses, it confers significantly reduction in attack accuracy while minimally impacting the model's worst case utility. Our approach can also be improved in the future via better control over the generator and specification of $\sim_I$ to better satisfy privacy advocates, regulators, and modelers together.

Acknowledgments. We thank Drs. Bressler, Liu (John Hopkins University (JHU) School of Medicine) and Delalibera (Eye Hospital, Brasilia, Brazil) for their help assessing images in Fig. 3. This work was funded by the JHU Institute for Assured Autonomy.

References

1. Abadi, M., et al.: Deep learning with differential privacy. In: Proceedings of the 2016 ACM SIGSAC Conference on Computer and Communications Security, pp. 308–318 (2016)
2. Bagdasaryan, E., Shmatikov, V.: Differential privacy has disparate impact on model accuracy. arXiv preprint arXiv:1905.12101 (2019)
3. Burlina, P., Freund, D.E., Dupas, B., Bressler, N.: Automatic screening of age-related macular degeneration and retinal abnormalities. In: 2011 Annual International Conference of the IEEE Engineering in Medicine and Biology Society, pp. 3962–3966. IEEE (2011)
4. Burlina, P.M., Joshi, N., Pekala, M., Pacheco, K.D., Freund, D.E., Bressler, N.M.: Automated grading of age-related macular degeneration from color fundus images using deep convolutional neural networks. JAMA Ophthalmol. **135**(11), 1170–1176 (2017)
5. Carlini, N., et al.: An attack on instahide: is private learning possible with instance encoding? arXiv preprint arXiv:2011.05315 (2020)
6. Carlini, N., et al.: Extracting training data from large language models. arXiv preprint arXiv:2012.07805 (2020)
7. Esteva, A.: Dermatologist-level classification of skin cancer with deep neural networks. Nature **542**(7639), 115–118 (2017)
8. EyePACS: Diabetic retinopathy detection (2015). Data retrieved from Kaggle. https://www.kaggle.com/c/diabetic-retinopathy-detection
9. Fredrikson, M., Lantz, E., Jha, S., Lin, S., Page, D., Ristenpart, T.: Privacy in pharmacogenetics: an end-to-end case study of personalized warfarin dosing. In: USENIX Security Symposium (2014)
10. Goodfellow, I., et al.: Generative adversarial nets. Adv. Neural. Inf. Process. Syst. **27**, 2672–2680 (2014)
11. Graves, A., Mohamed, A.R., Hinton, G.: Speech recognition with deep recurrent neural networks. In: 2013 IEEE International Conference on Acoustics, Speech and Signal Processing, pp. 6645–6649. IEEE (2013)
12. Gulshan, V., et al.: Development and validation of a deep learning algorithm for detection of diabetic retinopathy in retinal fundus photographs. JAMA **316**(22), 2402–2410 (2016)
13. He, K., Zhang, X., Ren, S., Sun, J.: Deep residual learning for image recognition. In: Proceedings of the IEEE Conference on Computer Vision and Pattern Recognition, pp. 770–778 (2016)
14. Heusel, M., Ramsauer, H., Unterthiner, T., Nessler, B., Hochreiter, S.: GANs trained by a two time-scale update rule converge to a local nash equilibrium (2018)

15. Huang, G., Liu, Z., Van Der Maaten, L., Weinberger, K.Q.: Densely connected convolutional networks. In: Proceedings of the IEEE Conference on Computer Vision and Pattern Recognition, pp. 4700–4708 (2017)
16. Jia, J., Salem, A., Backes, M., Zhang, Y., Gong, N.Z.: Memguard: defending against black-box membership inference attacks via adversarial examples. In: Proceedings of the 2019 ACM SIGSAC Conference on Computer and Communications Security, pp. 259–274 (2019)
17. Joshi, C.: Generative adversarial networks (GANs) for synthetic dataset generation with binary classes (2019). https://datasciencecampus.ons.gov.uk/projects/generative-adversarial-networks-gans-for-synthetic-dataset-generation-with-binary-classes
18. Kaissis, G.A., Makowski, M.R., Rückert, D., Braren, R.F.: Secure, privacy-preserving and federated machine learning in medical imaging. Nat. Mach. Intell. **2**(6), 305–311 (2020)
19. Karras, T., Aittala, M., Hellsten, J., Laine, S., Lehtinen, J., Aila, T.: Training generative adversarial networks with limited data. arXiv preprint arXiv:2006.06676 (2020)
20. Li, J., Li, N., Ribeiro, B.: Membership inference attacks and defenses in supervised learning via generalization gap. arXiv preprint arXiv:2002.12062 (2020)
21. Li, L., Verma, M., Nakashima, Y., Nagahara, H., Kawasaki, R.: IterNet: retinal image segmentation utilizing structural redundancy in vessel networks. In: The IEEE Winter Conference on Applications of Computer Vision (WACV), March 2020
22. Liu, K., Tan, B., Garg, S.: Subverting privacy-preserving GANs: hiding secrets in sanitized images (2020)
23. Nasr, M., Shokri, R., Houmansadr, A.: Machine learning with membership privacy using adversarial regularization. In: Proceedings of the 2018 ACM SIGSAC Conference on Computer and Communications Security, pp. 634–646 (2018)
24. Pekala, M., Joshi, N., Liu, T.A., Bressler, N.M., DeBuc, D.C., Burlina, P.: Deep learning based retinal OCT segmentation. Comput. Biol. Med. **114**, 103445 (2019)
25. Radford, A., Wu, J., Child, R., Luan, D., Amodei, D., Sutskever, I.: Language models are unsupervised multitask learners. OpenAI Blog **1**(8), 9 (2019)
26. Rogers, A., Kovaleva, O., Rumshisky, A.: A primer in BERTology: what we know about how BERT works. Trans. Assoc. Comput. Linguist. **8**, 842–866 (2021)
27. Salem, A., Zhang, Y., Humbert, M., Fritz, M., Backes, M.: ML-leaks: model and data independent membership inference attacks and defenses on machine learning models. In: Network and Distributed Systems Security Symposium 2019. Internet Society (2019)
28. Shokri, R., Stronati, M., Song, C., Shmatikov, V.: Membership inference attacks against machine learning models. In: 2017 IEEE Symposium on Security and Privacy (SP), pp. 3–18. IEEE (2017)
29. Ting, D.S.W., et al.: Artificial intelligence and deep learning in ophthalmology. Br. J. Ophthalmol. **103**(2), 167–175 (2019)
30. Ting, D.S., et al.: Deep learning in ophthalmology: the technical and clinical considerations. Prog. Retin. Eye Res. **72**, 100759 (2019)
31. Topol, E.J.: High-performance medicine: the convergence of human and artificial intelligence. Nat. Med. **25**(1), 44–56 (2019)
32. Vaswani, A., et al.: Attention is all you need. arXiv preprint arXiv:1706.03762 (2017)

33. Vizitiu, A., Niţă, C.I., Puiu, A., Suciu, C., Itu, L.M.: Towards privacy-preserving deep learning based medical imaging applications. In: 2019 IEEE International Symposium on Medical Measurements and Applications (MeMeA), pp. 1–6. IEEE (2019)
34. Yeom, S., Giacomelli, I., Fredrikson, M., Jha, S.: Privacy risk in machine learning: analyzing the connection to overfitting. In: 2018 IEEE 31st Computer Security Foundations Symposium (CSF), pp. 268–282. IEEE (2018)
35. Zhang, R., Isola, P., Efros, A.A., Shechtman, E., Wang, O.: The unreasonable effectiveness of deep features as a perceptual metric. In: CVPR (2018)
36. Zoph, B., Vasudevan, V., Shlens, J., Le, Q.V.: Learning transferable architectures for scalable image recognition. In: Proceedings of the IEEE Conference on Computer Vision and Pattern Recognition, pp. 8697–8710 (2018)

Author Index

Printed in the United States
by Baker & Taylor Publisher Services

Printed in the United States
by Baker & Taylor Publisher Services